O Level Physics for Cambridge Students: 2024 and 2025 Edition

Azhar ul Haque Sario

Copyright

First Printing, 2024

Azhar.sario@hotmail.co.uk

ORCID: https://orcid.org/0009-0004-8629-830X

Disclaimer: This book is free from AI use. The cover was designed in Microsoft Publisher. This book covers the complete syllabus for the Cambridge O Level Physics 2024-to-2025-year syllabus. It is the author's original work and has not been copied from other sources. It is intended as a supplementary resource for exam preparation.

Contents

Contents

Dynamics and Energetics

Measurement Techniques in Physics

Unlocking the Secrets of Measurement: A Journey Through Tapes, Rulers, and Micrometers

Imagine yourself as a detective, meticulously investigating the dimensions of the world around you. Your tools? Not a magnifying glass and fingerprint dust, but a humble tape measure, a trusty ruler, and the precision of a micrometer. Each tool holds a key to unlocking the secrets of size and scale, from the everyday to the extraordinary.

Tapes: The Flexible Detectives

Like a chameleon adapting to its surroundings, the tape measure takes on the form of its target. Cloth tapes, soft and pliable, gracefully measure the curves of a tailor's mannequin or the circumference of a towering oak. Steel tapes, rigid and strong, stand firm against the straight lines of a construction site or the vast expanse of a surveyor's landscape.

Tips from the Tape Measure Masters:

Tautness is Key: Imagine a tightrope walker maintaining perfect balance – a taut tape measure ensures accuracy, preventing the "sagging suspect" from distorting the truth.
Zero In on the Start: Every detective begins at the beginning. Aligning the tape's zero mark with your starting point sets the stage for a precise investigation.
Eye-Level Interrogation: Avoid the deceptive whispers of parallax errors by reading the measurement head-on, like a detective locking eyes with a suspect.
Material Matters: Just as a detective considers a suspect's background, consider the tape's material and thickness. These factors can subtly influence the outcome of your measurement.

Rulers: The Straight Investigators

Rulers, the steadfast guardians of straight lines, come in various forms. The standard ruler, a familiar friend, guides us through everyday measurements. The engineer's ruler, with its decimal precision, speaks the language of technical drawings and blueprints. And the metric ruler, a citizen of the global scientific community, communicates in the universal language of centimeters and millimeters.

Tips from the Ruler Experts:

Alignment is Everything: Like a detective lining up suspects in a police lineup, ensure the ruler's zero mark aligns perfectly with your starting point.
Mark Your Territory: A sharp pencil or marker becomes your detective's chalk, outlining the endpoint of your measurement with undeniable clarity.
The Parallax Paradox: Once again, avoid the deceptive whispers of parallax errors by reading the measurement head-on, maintaining a detective's unwavering gaze.

Micrometers: The Masters of Minute Detail

Enter the realm of the micrometer, where precision reigns supreme. These instruments, with their intricate mechanisms and delicate touch, unveil the hidden dimensions of the microscopic world. Outside micrometers embrace the external form, while inside micrometers delve into the depths of hidden cavities. Depth micrometers, like spelunkers exploring a cave, measure the depths of holes and recesses.

Tips from the Micrometer Maestros:

Cleanliness is Crucial: A detective wouldn't touch a crime scene with dirty hands. Similarly, clean the micrometer's anvil and spindle to ensure pristine measurements.
Gentle Persuasion: Like a detective coaxing a confession, close the micrometer gently on the object, using the ratchet to avoid excessive pressure.

Multiple Witnesses: Gather evidence from multiple vantage points by taking readings at different points on the object and averaging them for a comprehensive view.

Measuring Cylinders: The Liquid Investigators

Imagine a detective pouring over clues, not in a dusty file, but in a transparent cylinder. Measuring cylinders, the detectives of the liquid world, reveal the volumes of solutions, potions, and elixirs.

Tips from the Cylinder Connoisseurs:

The Right Tool for the Job: Choose a cylinder that matches the scale of your investigation, with a capacity slightly larger than the estimated volume of your liquid.
Stability is Essential: Just as a detective needs a steady hand to examine evidence, place the cylinder on a flat, stable surface for accurate readings.
The Meniscus Mystery: The meniscus, the curved surface of the liquid, holds a clue to the true volume. Read the measurement at its lowest point, like a detective focusing on the most telling detail.

Embrace the Detective Within

With these tools and techniques at your disposal, you become a measurement maestro, a detective of dimensions, uncovering the hidden secrets of size and scale. Whether you're measuring the length of a pencil, the volume of a potion, or the thickness of a metal sheet, embrace the precision and curiosity of a true investigator.

Tick-Tock, Goes the Clock: A Whimsical Journey Through Time Measurement

Time, that elusive river, flows onward with a steady current. But how do we capture its fleeting moments, those precious grains of sand slipping through our fingers? Fear not, fellow time travelers, for we have at our disposal ingenious contraptions, magical devices to measure time's passage!

Analog Clocks: Where Time Dances in Circles

Imagine a miniature stage where graceful hands waltz to the rhythm of seconds, minutes, and hours. The analog clock, with its charming face and twirling dancers, is a timeless classic.

The Dance of the Hands: Picture the second hand as a nimble ballerina, pirouetting with each passing second. The minute hand, a stately count, bows with every minute gone by. And the hour hand, the wise old sage, marks the grand passage of time.

Capturing Time's Footprints: To measure a time interval, we simply observe where the dancers are on the stage at the start of our event and where they end up when the curtain falls. The difference between these positions reveals the time that has danced by.

A Touch of Imperfection: Like any stage performance, the analog clock has its limitations. Our ballerina's swift movements can be a blur, making it tricky to measure those fleeting seconds with absolute precision. And let's face it, sometimes our eyes deceive us, leading to a slightly off-beat measurement.

Digital Timers: Time's Precise Reckoning

Enter the digital timer, a marvel of modern technology, where numbers flash with unwavering accuracy, leaving no room for doubt.

Stopwatches: The Timekeeper's Best Friend: These handy gadgets are like miniature race officials, capturing every millisecond with laser-sharp precision. With a simple click, we can start, stop, and reset the flow of time, recording those fleeting moments with utmost accuracy.

Timers: The Countdown to Adventure: Imagine a tiny hourglass, counting down the grains of time until something exciting happens. Whether it's a cake baking to perfection or a rocket launching into space, the timer keeps us on track, alerting us when the magic moment arrives.

The Perks of Precision: Digital timers are the champions of accuracy, banishing human error with their clear, unambiguous displays. They're the timekeepers we can truly rely on, whether we're timing a scientific experiment or simply boiling an egg.

Beyond the Everyday: Timekeeping for the Extraordinary

For those seeking to measure time with unparalleled accuracy, we venture into the realm of the extraordinary.

Atomic Clocks: Time's Ultimate Maestro: These incredible devices, with their atomic oscillations, are the guardians of time's true rhythm. They measure time with such precision that they lose less than a second every hundred million years!

High-Speed Cameras: Freezing Time's Fleeting Dance: Imagine a camera so fast that it can capture a hummingbird's wings mid-flight or a bullet frozen in time. These high-speed marvels allow us to measure incredibly short intervals, revealing the secrets of events that happen in the blink of an eye.

Data Loggers: The Time Chroniclers: These diligent recorders meticulously document events, marking each moment with a precise timestamp. They're the historians of the scientific world, preserving time's passage for analysis and discovery.

Measuring the World Around Us: Distance and Time's Dance

Time and distance are intertwined partners in a cosmic dance. To understand their relationship, we need to measure both with precision.

Rulers and Measuring Tapes: The Cartographers of Distance: These trusty tools help us map out the world, measuring everything from the length of a ladybug to the height of a skyscraper.

Vernier Calipers and Micrometers: The Masters of Miniature Measurement: When we need to measure the tiniest details, these precision instruments come to the rescue, revealing the subtle dimensions of the world around us.

Stopwatches and Digital Timers: The Timekeepers of Motion: Whether it's a ball rolling down a ramp or a pendulum swinging to and

for, these devices capture the rhythm of movement, allowing us to analyze the dance of distance and time.

The Pendulum's Swing: A Dance with Gravity

The pendulum, a simple weight suspended from a string, is a captivating timekeeper. Its rhythmic swing, governed by the invisible force of gravity, reveals a fascinating relationship between time, length, and the Earth's pull.

Unraveling the Pendulum's Secrets: By measuring the pendulum's period—the time it takes for one complete swing—we can unlock the secrets of gravity's influence.

A Dance Across the Globe: The pendulum's swing varies ever so slightly depending on its location on Earth. By carefully measuring these variations, we can map out the Earth's gravitational field, revealing hidden features beneath the surface.

Time's Tapestry: A Conclusion

From the simple elegance of an analog clock to the cutting-edge precision of atomic timekeeping, our journey through time measurement has revealed a world of fascinating tools and techniques. By understanding the interplay of time and distance, we can unlock the secrets of the universe and appreciate the intricate dance of the cosmos.

Scalars and Vectors: A Colorful Journey Through the Landscape of Quantities

Imagine yourself embarking on a thrilling adventure through the vibrant world of physics. As you traverse this landscape, you'll encounter fascinating creatures called "quantities," each possessing unique characteristics that shape their behavior. Some of these quantities, like shy scalars, are content with simply revealing their magnitude – a measure of their size or strength. Think of them as the introverts of the physics world, happy to tell you "How much" they possess, but not much else.

For instance, time, a wise old scalar, measures the duration of events, patiently ticking away without any sense of direction. Mass, a sturdy and reliable scalar, quantifies the amount of matter in an object, giving it a sense of presence in the universe. And then there's energy, the lifeblood of the universe, a scalar that fuels motion, change, and the very existence of everything around us.

But hold on tight, because this journey also introduces you to the more flamboyant vectors. These extroverted quantities are not only proud of their magnitude but also love to show off their direction, adding a dynamic twist to their personalities. They're like tiny arrows pointing the way, always eager to tell you “How much" and "where to."

Picture velocity, a speedy vector, zipping through space with a definite purpose. It's not just about how fast it's going, but also where it's headed. Then there's force, a powerful vector, pushing and pulling objects with its strength and direction, shaping the world around us. And let's not forget acceleration, a vector that captures the thrill of change in velocity, revealing not just how much an object is speeding up or slowing down, but also in which direction this change is happening.

Visualizing the Unseen

To truly appreciate these fascinating creatures, we need a way to visualize them. Scalars, being simple and straightforward, are easily represented by numbers. But vectors, with their dual nature, require a more expressive representation. Enter the arrow, a perfect symbol for a vector's magnitude and direction. The length of the arrow captures the vector's strength, while its direction points the way forward.

A Dance of Vectors

Vectors are not solitary beings; they love to interact and combine. Imagine them engaging in a graceful dance, where their movements create new patterns and possibilities. Vector addition is like a partnership where two vectors join forces, their arrows connecting head-to-tail, resulting in a new vector that represents their combined effect. Vector subtraction, on the other hand, is a playful tug-of-war, where vectors oppose each other, their arrows competing for dominance.

Real-World Encounters

This journey through the land of scalars and vectors isn't just a theoretical adventure; it's a practical exploration with real-world implications. Consider a pilot navigating an airplane through the skies. They rely on velocity vectors to account for both the plane's speed and the wind's direction, ensuring a safe and accurate flight path. Or think of an engineer designing a bridge, carefully analyzing the force vectors acting on the structure to ensure its stability and resilience against the weight of traffic and the forces of nature.

Unveiling the Secrets of the Universe

Scalars and vectors are not just abstract concepts; they are fundamental building blocks of the universe. They hold the key to understanding the intricate dance of particles, the forces that shape galaxies, and the very fabric of space and time. By grasping the essence of these quantities, we unlock a deeper understanding of the world around us and the mysteries that lie beyond.

So, embrace the adventure, explore the landscape of quantities, and let scalars and vectors guide you on a colorful journey through the fascinating world of physics.

Vectors: A Journey Through the Arrows of Reality

Imagine the world as a vast canvas, and every action, every movement, a stroke of vibrant color. But these strokes aren't just random splashes; they possess a distinct direction and intensity, much like arrows soaring through the air. These "arrows" are what we call vectors in the realm of physics, and they hold the key to understanding the intricate dance of motion, forces, and fields that shape our universe.

Unveiling the Essence of Vector Quantities

Let's embark on a journey to unravel the mysteries of these fundamental vector quantities:

Displacement: Imagine a playful puppy scampering around a park. While its paws trace a winding path, its displacement is the straight line connecting its starting point to its final resting spot, like an invisible arrow guiding its overall change in position. This concept is the backbone of navigation systems, ensuring you reach your destination efficiently, whether you're a lost puppy or a busy traveler.

Force: Picture a mighty tug-of-war, with teams pulling on a rope with all their might. The force they exert isn't just about strength; it's also about the direction of their pull. This push or pull, like an invisible hand guiding object, is crucial in fields like architecture and biomechanics, where understanding how structures withstand stress is paramount.

Weight: Ever felt the Earth's gentle embrace? That's weight, the force pulling you towards its core, like an invisible anchor keeping you grounded. This concept is vital in aerospace engineering, where precise calculations are needed to defy gravity and launch rockets into the vast expanse of space.

Velocity: Imagine a cheetah sprinting across the savanna, its speed and direction intertwined in a breathtaking display of velocity. This rate of change in displacement is not just about how fast something moves, but also where it's headed, like an arrow piercing through the air. Meteorologists rely on wind velocity to predict weather patterns and track storms, ensuring our safety in the face of nature's fury.

Acceleration: Picture a roller coaster, its velocity changing as it twists and turns, creating exhilarating thrills. This rate of change in velocity, whether in magnitude or direction, is like an invisible force propelling objects to new speeds and trajectories. Accelerometers in our smartphones capture these changes, enabling features like screen rotation and step counting, adding a touch of magic to our everyday lives.

Momentum: Imagine a bowling ball thundering down the lane, its mass and velocity combining to create an unstoppable force. This measure of how difficult it is to stop a moving object is like an invisible shield, protecting it from changes in motion. Momentum conservation is a fundamental principle in collision analysis, ensuring safety in fields like automotive design and particle physics.

Electric Field Strength: Imagine a charged object, its influence extending outwards like an invisible aura. This electric field strength determines the force it exerts on other charged objects, like an invisible web connecting them. This concept is crucial in designing electrical devices and understanding electromagnetic phenomena, powering our modern world.

Gravitational Field Strength: Imagine the Earth, its gravitational pull extending outwards like an invisible embrace. This gravitational field strength determines the weight of objects near its surface, like an invisible hand holding them close. Understanding this concept is essential for space exploration, satellite orbits, and studying celestial bodies, unraveling the mysteries of the cosmos.

Vectors in Action: Real-World Case Studies

Let's witness the power of vectors in action through captivating case studies:

Projectile Motion: Imagine a baseball soaring through the air, its trajectory a graceful arc. By breaking down its velocity into horizontal and vertical components, we can predict its range, maximum height, and time of flight, like a sports analyst predicting the outcome of a game.

Airplane Navigation: Imagine a pilot navigating through the skies, wind currents adding an extra layer of complexity. By adding the airplane's velocity vector to the wind velocity vector, they can determine the resultant velocity and adjust their heading accordingly, ensuring a safe and efficient journey.

Structural Analysis: Imagine a majestic bridge spanning a vast river, its structure withstanding immense forces. Engineers use vectors to represent these forces, ensuring the bridge's stability and safety, like a guardian protecting travelers on their journeys.

Conclusion: Vectors - The Guiding Arrows of Our Reality
Vectors are more than just abstract mathematical concepts; they are the guiding arrows that illuminate the intricate workings of our physical world. By understanding their properties and applications, we gain a deeper appreciation for the forces, motions, and fields that shape our everyday experiences and the vast expanse of the universe. So, let's embrace the power of vectors and continue exploring the fascinating world they reveal, one arrow at a time.

Motion

Buckle Up for a Joyride Through the World of Speed and Velocity!

Imagine a cheetah blazing across the savanna, a rocket soaring into space, or a dancer twirling gracefully across the stage. What do they all have in common? They're all in motion, baby! And to understand motion, we need to get a grip on two VIPs: speed and velocity.

Now, I know what you're thinking: "Aren't they the same thing?" Not quite! It's like this: speed is like a carefree spirit, just telling you how fast something's moving. Velocity, on the other hand, is a bit more uptight, like that friend who always needs to know exactly where you're going and how you'll get there.

Speed: The Need for Speed!

Think of speed as the adrenaline junkie of the physics world. It's all about how fast you can go, how quickly you devour distance. It's the thrill of a rollercoaster, the zoom of a sports car.

Instantaneous Speed: This is like checking your speedometer – it tells you your speed right now.
Average Speed: Imagine a road trip with crazy traffic and scenic detours. Your average speed is like summing up the whole journey – the fast bits, the slow bits, the "are we there yet?" bits.

Uniform Speed: This is like cruising on autopilot – steady and constant, covering the same distance in the same amount of time.
But what affects speed? Well, imagine trying to ride your bike uphill against a strong wind – that's friction and force in action! And don't forget about mass – a tiny scooter will zip along faster than a heavy truck with the same push.

Velocity: The GPS of Motion

Velocity is like your phone's navigation system – it's not just about how fast you're going, but also where you're headed. It's the difference between driving in circles (zero velocity, despite high speed!) and making a beeline for your destination.

Constant Velocity: Think of a train gliding along a straight track at a steady pace – that's constant velocity.
Changing Velocity: This is like navigating a winding road – you're constantly adjusting your speed and direction, and that's acceleration!
Terminal Velocity: Ever seen a skydiver gracefully floating down? That's terminal velocity – when the force of gravity and air resistance reach a truce, and the skydiver falls at a constant speed.
So, where does velocity come in handy? Everywhere! From guiding airplanes and ships to predicting the path of a curveball, velocity is the unsung hero of motion.

Real-World Adventures in Speed and Velocity

Let's take a look at some real-life examples to see these concepts in action:

The Marathon Runner: A marathon runner might have an impressive average speed, but if they end up back where they started, their average velocity is a big fat zero! It's all about that displacement, baby.
The Airplane: An airplane battling headwinds and tailwinds might have a wild ride, with its speed and velocity telling different stories.
Formula 1 Racing: These speed demons aren't just about raw speed, they're masters of velocity, taking those corners with precision and strategy.

Space Exploration: From launching rockets to slingshot ting around planets, space travel is a cosmic dance of speed and velocity calculations.

Beyond the Basics

But wait, there's more! The world of speed and velocity is full of exciting twists and turns:

Relative Velocity: Ever felt like you were standing still on a moving train? That's relative velocity – your speed depends on your point of view.
Angular Velocity: Think of a spinning top or a whirling dancer – that's angular velocity in action.

The Grand Finale

So, there you have it – a whirlwind tour of speed and velocity! From everyday life to the vastness of space, these concepts are the keys to unlocking the mysteries of motion. So next time you see a bird soaring, a car racing, or a ball flying, remember the dynamic duo of speed and velocity, working their magic behind the scenes.

Buckle Up for a Joyride Through the World of Motion!

Imagine a cheetah sprinting across the African savanna, a rocket blasting off into the cosmos, or a dancer gracefully twirling across a stage. What do all these have in common? Motion, of course! But how do we really understand and describe this motion? That's where our trusty trio – speed, velocity, and acceleration – come into play.

Speed: The Need for Speed!

Think of speed as the adrenaline junkie of the physics world. It's all about how fast something is going, the pure thrill of the motion! Imagine yourself cruising down the highway in a sleek sports car. Your speedometer tells you you're going 60 miles per hour – that's your speed! It doesn't matter if you're heading north, south, east, or west, speed only cares about how quickly you're devouring those miles.

Velocity: Speed Gets a Direction Upgrade!

Now, let's add a dash of purpose to our speed. Velocity is like speed with a built-in GPS. It not only tells you how fast you're going but also where you're headed. So, if you're cruising at 60 mph north, that's your velocity – a perfect blend of speed and direction. Think of it like this: speed is a carefree wanderer, while velocity is a determined explorer with a map and a mission!

Acceleration: The Shape-Shifter of Motion!

Hold on tight, because acceleration is where things get really interesting! It's all about change – how quickly your velocity is morphing. Are you slamming on the brakes? That's acceleration (in the negative direction, also known as deceleration). Are you flooring the gas pedal? That's acceleration too! Even if you're cruising at a constant speed but decide to take a sharp turn, you're still accelerating because your direction is changing. Acceleration is the master of transformation, always keeping things exciting!

Putting It All Together: The Motion Symphony!

Speed, velocity, and acceleration are like the instruments in an orchestra, each playing a unique part in the grand symphony of motion. Speed sets the tempo, velocity adds the melody, and acceleration brings in the dynamic changes that keep you on the edge of your seat.

Real-World Adventures in Motion:

Cheetah on the Hunt: Our cheetah friend relies on its incredible speed to catch its prey. But it's the cheetah's ability to accelerate quickly – to change its velocity rapidly – that truly makes it a master hunter.

Rocket to the Stars: A rocket launching into space needs tremendous speed to escape Earth's gravitational pull. But it's the constant acceleration, the continuous increase in velocity, that propels the rocket higher and higher.

Dancer in Motion: A dancer uses changes in speed and direction – changes in velocity – to create beautiful and expressive movements. And it's the acceleration, the smooth transitions between those movements, that gives the dance its fluidity and grace.

So, the next time you witness something in motion, remember our dynamic trio – speed, velocity, and acceleration. They are the keys to unlocking the secrets of movement, revealing the hidden patterns and rhythms that make our world so vibrant and alive!

Imagine this:

You're on a rollercoaster. Sometimes you're creeping upwards, the anticipation building with every click of the chain. Then, whoosh! You're hurtling down a steep drop, the wind whipping through your hair. That's non-uniform acceleration - your velocity, or speed and direction, is changing drastically from moment to moment.

Now, picture yourself cruising down an empty highway in a sleek sports car. You press the accelerator and feel a smooth, constant push as you gain speed. This is uniform acceleration - your velocity changes steadily, like a graceful dance.

Let's break it down:

Uniform Acceleration: Think of it as a steady heartbeat. The change in your speed happens at a constant rate. Imagine a train smoothly leaving a station, gaining speed in a predictable way. Plot this on a graph, and you'll see a straight line – elegant and predictable.

Non-Uniform Acceleration: This is more like a jazz solo – full of surprises and improvisation. A basketball player dribbling down the court, a leaf fluttering in the wind, or a car navigating busy city streets – these are all examples of non-uniform acceleration. On a graph, it's a wild, curvy line reflecting those unpredictable changes in speed and direction.

And what about slowing down?

That's simply acceleration in the opposite direction, like hitting the brakes on your bike. We call it deceleration, or sometimes even negative acceleration.

Think of it like this:

Imagine you're learning to drive. Mastering uniform acceleration is like getting comfortable with the gas pedal, smoothly increasing your speed. Non-uniform acceleration is like learning to navigate traffic, reacting to the world around you. And deceleration? That's all about those brakes!

Why does it matter?

Understanding these concepts helps us describe how objects move and interact in the world around us. It's the key to unlocking the secrets of motion, from the smallest particles to the largest stars.

So, the next time you're on a rollercoaster, in a car, or simply watching a ball drop, remember: you're witnessing the magic of acceleration in action.

Deciphering Distance-Time Graphs: A Visual Journey Through Motion

Imagine a lone hawk soaring above a vast desert, its keen eyes tracing the intricate paths of creatures below. Just like the hawk observes the movement of its prey, we can use distance-time graphs to visualize and understand the motion of objects. These graphs are like windows into the world of physics, revealing the hidden stories behind every movement.

Unveiling the Secrets of Motion: A Beginner's Guide to Distance-Time Graphs

Before we embark on our journey, let's equip ourselves with the basic tools of the trade:
The X-Axis (Time): This horizontal axis is the stage where time unfolds, measured in seconds, minutes, or even hours. It's like the ticking clock that governs the rhythm of motion.

The Y-Axis (Distance): This vertical axis measures the distance traveled by an object, typically in meters, kilometers, or miles. It's the yardstick that quantifies the extent of motion.
The Slope: The slope of the line on our graph is the key to unlocking the speed of an object. A steeper slope signifies a faster speed, while a gentler slope indicates a more leisurely pace.
The Shape: The shape of the line is the storyteller, revealing the nature of the motion, whether it's an object at rest, moving at a constant speed, accelerating, or decelerating.

1. The Stillness of Rest: A Horizontal Tale

Imagine a majestic tree standing tall and unyielding amidst a bustling city. Its position remains unchanged, just like an object at rest. On a distance-time graph, this stillness is depicted by a horizontal line, signifying zero speed.

Example: A car parked peacefully on a quiet street, its distance from a specific point remaining constant, is like a horizontal line on our graph.

Case Study: Picture a serene lake, its water level unwavering during a period of no rainfall or outflow. This tranquil scene translates into a horizontal line on our distance-time graph.

2. The Steady March of Constant Speed: A Linear Narrative

Imagine a train gliding along a straight track, covering equal distances in equal intervals of time. This steady motion is represented by a straight line with a non-zero slope on our graph. The steeper the slope, the faster the train's journey.

Example: A car cruising along a highway with its cruise control set, maintaining a constant speed, is like a straight line on our graph.

Case Study: Envision a determined marathon runner, maintaining a steady pace throughout the race. Their progress is reflected in a straight line on our distance-time graph.

3. The Thrill of Acceleration: A Curve of Increasing Speed

Imagine a roller coaster climbing a steep incline, its speed increasing with every passing moment. This exhilarating acceleration is depicted by a curve with an increasing slope on our graph. The steeper the curve, the greater the thrill.

Example: A car accelerating from a stop sign, gaining speed as it merges onto a busy road, is like a curve with an increasing slope on our graph.

Case Study: Picture a rocket launching into space, its speed increasing dramatically as it escapes Earth's gravitational pull. This powerful ascent is captured by a curve with a steepening slope on our graph.

4. The Gentle Slowdown of Deceleration: A Curve of Decreasing Speed

Imagine a car approaching a red light, gradually slowing down to a stop. This deceleration is represented by a curve with a decreasing slope on our graph. The steeper the curve, the more abrupt the slowdown.

Example: A ball thrown vertically upwards, slowing down as it ascends against gravity, is like a curve with a decreasing slope on our graph.

Case Study: Envision a parachutist descending from the sky, their speed gradually decreasing as they approach the ground. This gentle descent is depicted by a curve with a decreasing slope on our graph.

Beyond the Basics: Navigating the Complexities of Non-Uniform Motion

In the real world, motion is rarely as simple as constant speed or acceleration. Objects often change their speed and acceleration, creating a more intricate dance of motion. These variations are reflected in the shape of our distance-time graph, which may exhibit a combination of straight lines and curves.

Example: A car navigating through city traffic, accelerating, decelerating, and stopping at various points, is like a complex distance-time graph with varying slopes and shapes.

Case Study: Picture a bird soaring through the air, changing its speed and direction as it navigates the wind currents. This intricate flight pattern is captured by a complex distance-time graph with varying slopes and shapes.

Calculating Speed and Acceleration: Unlocking the Quantitative Secrets

Our distance-time graphs not only provide a visual representation of motion but also allow us to calculate the speed and acceleration of an object.

Speed: The slope of the line on our graph represents the speed of the object. We can calculate the speed by dividing the change in distance by the change in time.
Acceleration: The change in slope over time reveals the acceleration of the object. A steeper curve indicates a higher acceleration, while a gentler curve indicates a lower acceleration.

Conclusion: Mastering the Art of Motion Analysis

Distance-time graphs are powerful tools that allow us to visualize, analyze, and understand the motion of objects. By deciphering the shapes and slopes of these graphs, we can unlock the secrets of speed, acceleration, and the intricate dance of motion.

This comprehensive guide has equipped you with the knowledge to interpret distance-time graphs and extract meaningful information about an object's motion. So, embrace the power of these visual tools and embark on your own journey of motion analysis, unraveling the mysteries of movement in the world around you.

Imagine this:

Instead of a boring old book on a table, picture a mischievous squirrel perched on a branch, perfectly still, plotting its next daring leap. Its speed-time graph? A flat line, as straight as the squirrel's cunning gaze.

Now, envision a cheetah, the ultimate speedster, streaking across the savanna in hot pursuit of its prey. Its speed-time graph? A high-flying line, parallel to the time axis, mirroring the cheetah's unwavering velocity.

Feeling the acceleration?

Picture a skateboarder launching themselves down a ramp, gaining speed with every passing second. Their speed-time graph? A line that climbs higher and higher, reflecting their exhilarating acceleration.

But what about a rollercoaster, with its twists, turns, and sudden drops? Its speed-time graph would be a work of art, a dynamic curve capturing every exhilarating moment.

Let's dive deeper into some real-life adventures:

The Plunging Diver: Imagine a diver leaping from a towering cliff, gravity pulling them downwards. Their speed-time graph? A line that swoops downwards like the diver themselves, showcasing the relentless acceleration of their freefall.

The Soaring Rocket: Picture a rocket blasting off into the cosmos, its engines roaring with power. Its speed-time graph? A line that shoots upwards like a comet, mirroring the rocket's relentless climb towards the stars.

The City Bus: Envision a bus navigating the bustling streets, stopping and starting, accelerating and decelerating. Its speed-time graph? A series of peaks and valleys, a visual symphony of urban motion.
The Sneaky Snail: Can you sketch a speed-time graph for a snail embarking on its epic journey across a garden?
The Daring Skydiver: How would you depict the motion of a skydiver, from freefall to parachute deployment, on a speed-time graph?

The Bouncing Ball: Can you capture the rhythmic motion of a bouncing ball with a series of curves and lines?
Remember:

Speed-time graphs are more than just lines on paper; they're stories waiting to be told. With a dash of imagination, you can unlock the secrets of motion and transform these graphs into captivating narratives.

So, embrace your creativity, explore the world around you, and let the lines on a speed-time graph guide you on a journey of discovery!

Imagine this:

You're standing at the edge of a cliff, heart pounding, about to take a leap of faith. Not a suicidal one, mind you, but a thrilling skydives! As you step off the edge, the world seems to fall away, and you're in freefall. The ground rushes towards you, faster and faster.

This exhilarating experience perfectly illustrates the concept of freefall acceleration. You see, Earth has this invisible pull called gravity, and it's constantly tugging on you, like an over-eager puppy with your favorite toy. This pull causes you to accelerate downwards, gaining speed every single second. Scientists have a fancy name for this acceleration – they call it 'g', and it's roughly 9.8 meters per second squared. That means for every second you're falling, your speed increases by about 9.8 meters per second. Whoa!

Now, let's say you want to track your skydiving adventure. You could use a distance-time graph, which is like a map of your fall. It shows how far you've fallen at different points in time. The steeper the line on the graph, the faster you're falling. It's like comparing a leisurely stroll to a frantic sprint – the sprint would have a much steeper line!

But here's the cool part: you can actually calculate your speed at any point during your fall just by looking at this graph. It's like having a speedometer built into your map! You simply pick two points on the line, figure out how much distance and time you've covered between those points, and divide the distance by the time. Voilà, you've got your speed!

Think of it like this: imagine two snapshots from your skydiving video. In the first snapshot, you're 10 meters from the plane after 2 seconds. In the second snapshot, you're a daredevil 30 meters away after 4 seconds. To find your average speed during that time, you'd subtract the distances (30 - 10 = 20 meters) and subtract the times (4 - 2 = 2 seconds). Then, divide the distance by the time (20 / 2 = 10 meters per second). That's your speed!

Of course, freefall isn't just for skydivers. It's everywhere! Think of a basketball soaring towards the hoop, a raindrop plummeting from the sky, or even a playful puppy leaping for a treat. They're all experiencing the effects of gravity and acceleration.

And distance-time graphs? They're not just for tracking falling objects either. You can use them to analyze the motion of a car speeding down the highway, a snail slowly creeping across a leaf, or even the progress of a particularly exciting race!

So, the next time you see an object in motion, remember the thrill of the skydive, the pull of gravity, and the power of a simple graph to unlock the secrets of speed and motion. It's a whole new way of looking at the world around you!

Imagine a cheetah on the savanna, its sleek body a blur as it chases a gazelle. That's constant speed in action! A speed-time graph would show this as a flat line, like the horizon stretching across the plains.

Now picture a falcon diving from the sky. It starts slow, then plummets faster and faster towards its prey. That's constant acceleration! On our graph, we'd see a line sloping upwards, like the falcon's breathtaking descent.

But what about a squirrel scampering through the branches? Darting, pausing, leaping – its speed is always changing. This is non-uniform acceleration, a wild, squiggly line on our graph, full of surprises!

Think of the area under the graph as the distance covered. If our cheetah sprints for ten seconds at top speed, the rectangle under the line tells us how far it ran. It's like measuring the cheetah's paw prints across the savanna!

The slope of the line? That's acceleration. Steeper means faster acceleration, like our falcon gaining speed. A downward slope? That's deceleration, like a car hitting the brakes.

Speed-time graphs are like secret codes, revealing the hidden stories of motion. They help us understand everything from the graceful arc of a thrown ball to the thrilling twists and turns of a roller coaster.

Want to know how fast a rocket accelerates? Or how far a dancer leaps across the stage? A speed-time graph holds the answers! It's a window into the world of movement, where lines and curves come alive with the stories of speed and time.

So next time you see an object in motion, imagine its speed-time graph. It's a hidden language, waiting to be decoded. And once you understand it, the world of motion will never look the same!

Mass and weight

Imagine a Universe Made of Marshmallows

Okay, let's ditch the textbooks and dive headfirst into the squishy world of mass. Forget the stuffy definitions and rigid formulas. Instead, picture a universe made entirely of marshmallows.

More Marshmallows, More Mass

Imagine holding a single marshmallow in your hand. It's light, fluffy, and easy to move around. Now, picture a giant bag overflowing with marshmallows. It's heavier, harder to move, and definitely more substantial. That, my friend, is the essence of mass! The more marshmallows you have, the more mass you have. Simple, right?

The "Couch Potato" Effect

Now, let's talk about inertia. Think of it as the "couch potato" effect. The more marshmallows you pile onto your couch (more mass), the harder it is to get that couch moving (more inertia). It's like that feeling of being glued to the sofa after a Thanksgiving feast. The more you've indulged, the less likely you are to spring up and do the dishes.

Why You Weigh Less on the Moon (But Still Need a Diet)

Ever dreamed of bouncing around on the moon? Well, here's a fun fact: you'd weigh less on the moon, but your mass would stay the same. Confused? Think of it this way: you're still made of the same number of "marshmallows" (mass), but the moon's gravity is weaker, so it pulls on those marshmallows with less force (weight). So, go ahead, take that giant leap for mankind! Just don't expect to lose any marshmallow mass in the process.

Einstein's Marshmallow Surprise

Now, for the mind-blowing part. Einstein, the ultimate cosmic chef, discovered that mass and energy are like two sides of the same marshmallow coin. They can transform into each other! It's like magically turning a handful of marshmallows into a burst of pure energy (think exploding supernovae!). Or, conversely, taking a blast of energy and condensing it into a brand-new marshmallow. Pretty wild, huh?

Marshmallows and the Mysteries of the Universe

From the Higgs boson (a tiny, elusive marshmallow that gives other particles mass) to black holes (cosmic vacuum cleaners that gobble up marshmallows like there's no tomorrow), mass plays a starring role in the grand cosmic drama. It shapes galaxies, fuels stars, and even dictates how the universe expands.

So, the next time you encounter a marshmallow, take a moment to appreciate its hidden depths. It's not just a tasty treat; it's a window into the fundamental fabric of reality.

Unlocking the Mysteries of Mass: A Journey Through Time and Technology

Imagine holding a feather and a rock in your hands. They feel vastly different, right? This difference, my friend, is due to mass – the amount of matter packed inside an object. Unlike weight, which changes depending on where you are in the universe (imagine bouncing around on the moon!), mass stays constant. It's a fundamental property, a cosmic fingerprint of sorts.

Now, how do we measure this elusive property? Let's embark on a journey through time and technology, exploring two fascinating tools:

1. The Beam Balance: A Classic Tale of Equilibrium

Picture an ancient marketplace, bustling with merchants and traders. In the midst of the chaos, a simple yet ingenious device holds court – the beam balance. This elegant contraption, with its two pans suspended from a central pivot, has been used for centuries to ensure fair trade.

But how does it work? Imagine a seesaw perfectly balanced with two friends of equal weight. Now, if one friend suddenly gains a few pounds, the seesaw tilts. To restore balance, you'd need to add weight to the other side. The beam balance operates on the same principle – the principle of moments.

Balancing Act: You place the object of unknown mass on one pan and add standard weights to the other until the beam balances perfectly. Voilà! The mass of the object equals the mass of the standard weights. It's like a cosmic dance of equilibrium!

Pros: Simple, durable, and requires no power source. It's the perfect tool for comparing masses and has a timeless charm.

Cons: Not as precise as its modern counterpart and can be a bit slow. It's also sensitive to environmental factors like wind and vibrations.

2. The Electronic Balance: A Digital Dance of Precision

Fast forward to the modern age, where technology reigns supreme. Enter the electronic balance, a sleek and sophisticated device that measures mass with incredible accuracy.

Electromagnetic Magic: This marvel of engineering uses an electromagnetic force to counterbalance the weight of the object. It's like an invisible hand delicately holding the object in place. The amount of force required is directly proportional to the mass, which is displayed on a digital screen.

Pros: Highly precise, fast, and easy to read. It's like having a mini-laboratory at your fingertips!

Cons: Requires a power source and can be more expensive than a beam balance.

Beyond the Balance: Other Ways to Weigh the Universe

While beam balances and electronic balances are the workhorses of mass measurement, there are other intriguing methods:

Inertia Balance: This clever device measures mass based on an object's resistance to acceleration. It's like trying to push a heavy box versus a light one – the heavier the object, the harder it is to get moving.

Mass Spectrometer: This powerful tool is used to identify and quantify different substances by measuring the mass-to-charge ratio of ions. It's like a molecular detective, uncovering the hidden secrets of matter.

Gravitational Methods: Astronomers use the gravitational attraction between objects to measure the mass of celestial bodies. It's a mind-boggling concept, using the very fabric of the universe as a weighing scale!

The Enduring Quest to Understand Mass

From ancient marketplaces to cutting-edge laboratories, the quest to understand and measure mass has driven scientific inquiry for centuries. Whether you're a student, a scientist, or simply curious about the world around you, I hope this journey has shed light on this fundamental property of matter. So, the next time you hold an object in your hand, take a moment to appreciate the intricate dance of atoms that gives it mass – a property that connects us to the vast universe.

Unraveling the Mysteries of Weight with a Force Meter: A Hands-On Adventure!

Ever wondered how much force that apple exerted on Newton's head? Or how scientists measure the weight of a rocket before it blasts off to explore the cosmos? Well, get ready to dive into the fascinating world of force meters, where we'll unlock the secrets of weight, gravity, and the tools that help us understand these fundamental forces!

1. Force Meters: Your Pocket-Sized Gravity Detective

Imagine a spring with a superpower: the ability to reveal the weight of objects! That's essentially what a force meter is – a clever contraption also known as a newton meter or spring scale. It's like having a mini detective kit to uncover the hidden forces around us.

But hold on! Weight isn't the same as mass, even though we often use the terms interchangeably. Mass is like a stubborn friend who always stays the same, no matter where they are. Weight, on the other hand, is a bit of a chameleon, changing depending on the gravitational pull. It's the force that makes you feel lighter on the moon and heavier on Jupiter (imagine trying to do a cartwheel there!).

2. Weighing In: A Step-by-Step Guide to Force Meter Mastery

Ready to become a weight-measuring wizard? Here's your secret recipe:

a) Choosing Your Weapon:

Capacity: Just like you wouldn't use a teaspoon to measure a gallon of milk, you need a force meter that can handle the weight of your object. Overloading it is like trying to squeeze an elephant into a phone booth – not a good idea!

Calibration: Think of calibration as tuning your instrument, like a musician tuning their guitar before a concert. It ensures your force meter is playing the right tune and giving you accurate readings.

b) Preparing for the Weigh-In:

Zeroing In: If your force meter has a zero adjustment, make sure it reads zero when nothing is attached. It's like resetting the scale before stepping on it.
Gravity Check: Hold the force meter vertically, allowing gravity to work its magic directly on the spring.

c) Hooking Your Catch:

Secure Attachment: Gently attach your object to the force meter's hook, making sure it's secure. We don't want any surprise "flying objects" during our experiment!

d) The Moment of Truth:

Steady as She Goes: Let the object settle down and the spring find its balance. Any swinging or shaking will throw off your readings.
Eye Spy: Read the measurement at eye level to avoid any optical illusions. It's like lining up the perfect shot in a game of billiards.

e) Recording the Victory:

Newton's Calling: Remember, weight is measured in newtons (N), named after the legendary Sir Isaac Newton himself.
Precision Matters: Record your measurement with the right level of detail, depending on the force meter's scale.

3. Gravitational Field Strength: The Invisible Force That Keeps Us Grounded

Imagine an invisible web surrounding Earth, pulling everything towards its center. That's the gravitational field, and its strength determines how strongly objects are pulled towards the planet. It's like Earth's way of saying, "Stay close, I've got you!"

The formula $g = W / m$ is your key to understanding this invisible force. It tells you how much force (weight) is exerted on each unit of mass.

4. Free Falling and Gravity's Grip

Ever jumped off a diving board and felt that exhilarating rush as you accelerate towards the water? That's free fall, and it's intimately connected to gravitational field strength. The acceleration you experience is equal to the strength of the gravitational field. It's like gravity is saying, "I'm in control, and I'm taking you for a ride!"

5. Gravity's Mood Swings: Factors Affecting its Strength

Gravity isn't always the same. It can be a bit moody, changing its strength depending on:

Distance: The farther you are from Earth's center, the weaker its pull. It's like a long-distance relationship – the further apart you are, the weaker the connection.
Mass: Massive objects have a stronger gravitational pull. Think of the Sun, holding all the planets in its orbit like a cosmic dance.
Mass Distribution: Even the way mass is spread out within an object can affect gravity's strength.

6. Real-World Adventures with Force Meters

The Curious Case of the Book: Grab a force meter and your favorite book. Attach the book, let it hang, and voilà! You've just measured its weight.
Moonwalking with Newton: Imagine you're an astronaut on the moon, where gravity is much weaker. Using a force meter and the formula $g = W / m$, you can calculate the gravitational field strength on the lunar surface.
Rocket Science 101: Before a rocket launches, engineers use force sensors (similar to force meters) to measure its weight. This helps them calculate the thrust needed to escape Earth's gravitational clutches.

7. The Grand Finale: Weighing in on the Importance of Force Meters

Force meters are more than just simple tools; they're gateways to understanding the forces that shape our universe. By measuring weight and exploring the mysteries of gravity, we gain a deeper appreciation for the delicate balance that keeps us grounded and allows us to explore the cosmos. So, embrace your inner scientist, grab a force meter, and embark on a journey of discovery!

Imagine the universe as a vast, invisible spiderweb, glistening with dewdrops of gravity. Each dewdrop, a celestial body, creates a dip in the web, a gravitational field that tugs at anything that ventures too close. The bigger the dewdrop, the deeper the dip, the stronger the pull.

Think of Earth, our home, as a particularly plump dewdrop. We, its inhabitants, are like tiny insects clinging to its surface, held fast by its gravitational embrace. This invisible force, this cosmic glue, is what keeps our feet firmly planted on the ground and prevents us from floating off into the starry abyss.

Picture yourself as an astronaut, venturing out into the cosmos. As you drift away from Earth, you feel the pull of its gravity weaken, like a rubber band stretching thinner and thinner. You enter a state of weightlessness, a sensation of freedom and detachment from the world below.

Now, imagine yourself hurtling towards a black hole, a cosmic sinkhole of unimaginable density. Its gravitational pull is so intense that it warps the very fabric of space and time, like a bowling ball dropped onto a trampoline. Light itself cannot escape its clutches, hence its name.

Gravity is the architect of the universe, the invisible hand that sculpts galaxies and guides the dance of celestial bodies. It is a force of both creation and destruction, pulling together gas and dust to form stars, yet also capable of crushing them into oblivion.
This is the magic of gravitational fields, a cosmic ballet of attraction and influence. It is a force that connects us all, from the smallest atom

to the largest galaxy, a reminder that we are all part of a grand, interconnected cosmic web.
So, the next time you feel the ground beneath your feet, take a moment to appreciate the invisible force that keeps you tethered to this planet. It is a force that has shaped our world, our solar system, and the universe itself.

Density

Density: A Journey into the Heart of Matter

Imagine yourself cradling a fluffy feather in one hand and a hefty rock in the other. They're roughly the same size, yet the rock feels significantly heavier. This difference in weight, despite their similar volumes, boils down to a fascinating property called density.

Density: Unpacking the Mass-Volume Relationship

In essence, density tells us how tightly matter is packed together.

Think of it as a measure of how much "stuff" is crammed into a given space. The denser an object, the more mass it packs into a specific volume.

The Equation: A Mathematical Lens

To calculate density, we use a simple yet powerful equation:

Density (ρ) = Mass (m) / Volume (V)

Where:

ρ (rho) represents density, typically measured in kilograms per cubic meter (kg/m^3) or grams per cubic centimeter (g/cm^3).
m represents mass, typically measured in kilograms (kg) or grams (g).
V represents volume, typically measured in cubic meters (m^3) or cubic centimeters (cm^3).
This equation reveals a fundamental relationship: density is directly proportional to mass and inversely proportional to volume. In simpler

terms, if you increase the mass of an object while keeping its volume constant, its density increases. Conversely, if you increase the volume while keeping the mass constant, the density decreases.

Unveiling Density: A Tale of Three States

The method for determining density varies depending on the state of matter (solid, liquid, or gas) and the shape of the object. Let's explore how to determine the density of liquids and solids:

a) Liquids: Diving into Fluidity

To determine the density of a liquid, follow these steps:

Measure the Mass:

Weigh an empty container (e.g., a beaker) using a balance.
Pour the liquid into the container and measure the combined mass.
Subtract the container's mass from the combined mass to obtain the liquid's mass.

Measure the Volume:

If using a graduated cylinder or volumetric flask, read the volume directly from the markings.
If using a container without precise volume markings, transfer the liquid into a graduated cylinder for measurement.

Calculate the Density:

Divide the liquid's mass by its volume using the density equation: $\rho = m/V$.

Example:

A beaker weighs 25.0 g. When filled with 50.0 mL of ethanol, the combined mass is 64.8 g. What is the density of ethanol?

Mass of ethanol = 64.8 g - 25.0 g = 39.8 g
Volume of ethanol = 50.0 mL

Density of ethanol = 39.8 g / 50.0 mL = 0.796 g/mL

b) Regularly Shaped Solids: Embracing Geometry

For regularly shaped solids like cubes, rectangular prisms, spheres, and cylinders, follow these steps:

Measure the Mass:

Use a balance to measure the mass of the solid object.

Measure the Volume:

Use appropriate geometric formulas to calculate the volume based on the object's dimensions.
Cube: $V = side^3$
Rectangular Prism: V = length x width x height
Sphere: $V = (4/3)\ \pi r^3$ (where r is the radius)
Cylinder: $V = \pi r^2 h$ (where r is the radius and h are the height)
Use a ruler, caliper, or other measuring tool to accurately measure the dimensions.

Calculate the Density:

Divide the solid's mass by its volume using the density equation: $\rho = m/V$.

Example:

A cube of aluminum has a side length of 2.5 cm and a mass of 42.2 g. What is the density of aluminum?

Volume of aluminum cube = 2.5 cm x 2.5 cm x 2.5 cm = 15.625 cm^3
Density of aluminum = 42.2 g / 15.625 cm^3 = 2.70 g/cm^3

c) Irregularly Shaped Solids: The Art of Displacement

For irregularly shaped solids, we use the water displacement method:

Measure the Mass:

Use a balance to measure the mass of the solid object.

Measure the Volume by Displacement:

Fill a graduated cylinder or other container with a known volume of water. Record the initial water level.
Carefully submerge the solid object in the water, ensuring it's fully submerged and no air bubbles are trapped.
Record the new water level.
The difference between the initial and final water levels represents the volume of the solid.

Calculate the Density:

Divide the solid's mass by its volume (determined by displacement) using the density equation: $\rho = m/V$.

Example:

An irregularly shaped piece of rock has a mass of 150.0 g. When placed in a graduated cylinder containing 50.0 mL of water, the water level rises to 78.0 ML. What is the density of the rock?

Volume of rock = 78.0 mL - 50.0 mL = 28.0 mL
Density of rock = 150.0 g / 28.0 mL = 5.36 g/mL

Important Considerations:

Accuracy: The accuracy of density measurements depends on the precision of the instruments used. Use the most precise instruments available and take multiple measurements to minimize errors.
Temperature: Density can be affected by temperature, especially for liquids and gases. Record the temperature at which the measurements were taken.

Units: Ensure consistent units for mass and volume when calculating density.
Buoyancy: For objects that float, use a method that accounts for buoyancy.

Density in Action: Real-World Applications

Forensic Science: Density measurements help identify unknown substances, such as drugs, fibers, and glass fragments.
Gemology: Gemologists use density to identify gemstones and distinguish them from imitations.
Oceanography: Oceanographers study seawater density to understand ocean currents and marine life distribution.
Material Science: Engineers use density to select appropriate materials for specific applications, such as lightweight yet strong materials for aircraft construction.

Conclusion: Density's Enduring Significance

Density, a fundamental property of matter, plays a pivotal role in various scientific and engineering fields. Understanding how to measure and calculate density is essential for a wide range of applications, from identifying unknown substances to designing new materials. By carefully following the methods described above, you can accurately determine the density of liquids, regularly shaped solids, and irregularly shaped solids.

Forces

Forces: A Captivating Journey into the Unseen Movers of Our World

Imagine a world without motion, a universe frozen in stillness. No swirling galaxies, no rushing rivers, not even the gentle flutter of a butterfly's wings. It's hard to fathom, isn't it? That's because forces, those invisible agents of change, are constantly at play, shaping our dynamic reality. They're the puppet masters of the cosmos, the choreographers of existence, and today, we embark on a captivating journey to unravel their mysteries.

1. Unveiling the Diverse Personalities of Forces

Forces come in a variety of guises, each with its unique characteristics and influence on the world around us. Let's meet some of these fascinating characters:

a) Weight (The Earth's Loving Embrace)

Picture yourself standing on the Earth. You feel a gentle pull downwards, a comforting reminder that you're anchored to this planet. That's weight, the force of gravity drawing you towards the Earth's core. It's like the Earth's way of saying, "Stay a while, you're welcome here."

Formula: Weight (W) = mass (m) x acceleration due to gravity (g)
Direction: Always points downwards, towards the Earth's warm embrace.
Example: A playful puppy, bounding with joy, experiences weight as it leaps and frolics.
Case Study: Architects and engineers, like master builders, must carefully consider the weight of a skyscraper to ensure it stands tall and proud against the relentless tug of gravity.

b) Friction (The Ever-Present Resistance)

Imagine sliding a book across a table. It eventually comes to a halt, doesn't it? That's friction, the force that opposes motion, always trying to slow things down. It's like a mischievous imp, playfully hindering movement.

Types: Static friction (resisting initial motion) and kinetic friction (opposing ongoing motion).
Factors: Roughness of surfaces (think sandpaper versus silk) and the force pressing them together.
Example: A skilled rock climber relies on friction between their shoes and the rock face to defy gravity and ascend seemingly impossible heights.
Case Study: Brake pads in a car utilize friction to convert kinetic energy into heat, slowing the vehicle down and ensuring safety.

c) Drag and Air Resistance (The Invisible Hand that Shapes Speed)

Imagine a bird soaring through the sky. It encounters air resistance, a type of drag that opposes its motion. It's like an invisible hand gently caressing the bird, shaping its flight path.

Factors: The bird's streamlined shape, its speed, and the density of the air.
Example: A graceful dolphin leaping out of the water experiences drag as it breaks through the surface, creating a mesmerizing arc.
Case Study: Engineers designing high-speed trains carefully consider air resistance to minimize energy consumption and maximize efficiency.

d) Tension (The Force that Binds)

Imagine a tightrope walker gracefully balancing on a high wire. The rope is taut, experiencing tension, the force that transmits through it. It's like an invisible thread connecting the walker to the supports, keeping them aloft.

Direction: Acts along the rope, pulling equally on both ends.
Example: A guitar string, when plucked, vibrates due to tension, creating beautiful melodies.
Case Study: Suspension bridges, marvels of engineering, rely on the immense tension in their cables to support the weight of the deck and the vehicles that traverse it.

e) Electrostatic Force (The Dance of Attraction and Repulsion)

Imagine rubbing a balloon against your hair. It magically sticks to the wall! That's electrostatic force, the interaction between charged objects. It's like a captivating dance of attraction and repulsion.

Types: Opposite charges attract (like lovers drawn together), while like charges repel (like rivals pushing each other away).
Example: Lightning, a dramatic display of nature's power, is a massive electrostatic discharge.
Case Study: Photocopiers and laser printers utilize electrostatic forces to attract toner particles to the paper, creating sharp images and text.

f) Magnetic Force (The Invisible Force Field)

Imagine a compass needle pointing north, guided by the Earth's magnetic field. That's magnetic force, the force exerted by magnets on each other and magnetic materials. It's like an invisible force field, influencing the motion of objects within its realm.

Types: Opposite poles (north and south) attract, while like poles repel.
Example: Maglev trains, floating above the tracks, utilize magnetic levitation to achieve incredibly high speeds.
Case Study: MRI machines, powerful tools in medical diagnosis, use strong magnetic fields to generate detailed images of the human body's inner workings.

g) Thrust (The Driving Force of Progress)

Imagine a rocket launching into space, propelled by a powerful thrust. It's the force that pushes or pulls an object forward, overcoming resistance and enabling motion. It's like an engine of progress, driving us forward.

Direction: Opposite to the direction of motion.
Example: A jet engine, roaring with power, generates thrust to propel an airplane through the sky.
Case Study: Submarines utilize thrust to navigate the depths of the ocean, exploring the hidden wonders beneath the waves.

h) Contact Force (The Force of Touch)

Imagine pushing a shopping cart down the aisle. You're exerting a contact force, a force that requires direct contact between objects. It's the force of touch, the interaction between two surfaces.

Example: A gentle breeze caressing your face is a contact force, as is the satisfying thud of a basketball hitting the court.
Case Study: Robotics engineers carefully consider contact forces when designing robots that interact with their environment, ensuring they can grasp objects, navigate obstacles, and perform delicate tasks.

2. Mapping the Forces: Free-Body Diagrams

To truly understand the interplay of forces, we use free-body diagrams, visual representations that isolate an object and depict the forces acting upon it. They're like maps, guiding us through the intricate landscape of forces.

Steps to Draw a Free-Body Diagram:

Isolate the object: Focus on the object of interest.
Identify the forces: Consider all the forces acting on the object.
Represent as vectors: Draw arrows to represent each force, with length indicating magnitude and direction indicating the force's line of action.
Label the forces: Clearly label each force (e.g., W for weight, F and _f for friction).

Examples:

A book resting on a table: Weight downwards, normal force upwards.
A skydiver in freefall: Weight downwards, air resistance upwards.
A car accelerating: Thrust forward, friction backward, weight downwards, normal force upwards.
Case Study: Crash investigators use free-body diagrams to reconstruct accidents, analyzing the forces involved to determine the cause and contributing factors.

Conclusion: Forces, the Architects of Our Dynamic Universe

Forces are the unsung heroes of our universe, the silent conductors orchestrating the symphony of motion. They shape our world, from the grand dance of celestial bodies to the intricate movements of our own bodies. By understanding the diverse personalities of forces and their intricate interplay, we gain a deeper appreciation for the dynamic beauty of our universe.

Imagine a lazy dog. It's sprawled out on the living room rug, perfectly content to stay there for the rest of eternity. That's inertia in a nutshell – the tendency of things to just keep doing what they're already doing. Whether it's napping on a comfy rug or zooming through the vast emptiness of space, objects have a stubborn streak when it comes to changing their motion.

Newton's First Law of Motion, aka the "Law of Inertia," is like the universe's ultimate couch potato rule: "If I'm at rest, I'm staying at rest. And if I'm moving, I'm going to keep moving in a straight line at the same speed... unless someone or something forces me to change."

Think of it like this:

The stubborn coffee cup: Your coffee cup sitting on your desk isn't going to suddenly launch itself across the room unless you give it a push (or your cat gets some crazy ideas).
The runaway shopping cart: That shopping cart you're pushing down the aisle won't magically stop unless you apply the brakes (or crash into a display of pickles).
The cosmic cruise control: Picture a spaceship gliding through the frictionless vacuum of space. It'll keep sailing along at the same speed and direction forever unless it fires its engines or bumps into a rogue asteroid.
But what about forces? Well, they're the ultimate party crashers! A force can be a push, a pull, a nudge, or a slam – anything that has the power to change an object's motion.

Forces are like the remotes of the universe:

"Fast forward!": A force in the same direction as the motion speeds things up. Think of pushing a swing to make it go higher.
"Rewind!": A force in the opposite direction slows things down. Like when you catch a ball, bringing it to a stop.
"Turn left!": A force at an angle changes the direction. Imagine hitting a tennis ball with your racket, sending it flying across the court.
The bigger the object, the bigger the inertia. It's much easier to push a toy car than a real one, right? That's because the real car has more mass, and therefore more inertia – it's more resistant to changes in its motion.

Here's where things get really cool:

Seatbelts are lifesavers because of inertia. When a car slams on the brakes, your body wants to keep moving forward (thanks, inertia!). The seatbelt provides a force to stop you, preventing you from becoming a human projectile.
Shaking a ketchup bottle is a battle against inertia. That stubborn ketchup is at rest inside the bottle, and it takes a good shake (a force!) to overcome its inertia and get it moving.
Space travel is a testament to inertia. Once a spacecraft reaches its cruising speed in space, it can coast along for incredible distances without much fuel. Inertia keeps it going!
Newton's First Law is like a key that unlocks a deeper understanding of how the universe works. It's a fundamental principle that helps explain everything from the motion of planets to the physics of everyday objects. It's a reminder that objects have a natural tendency to resist change, and it takes a force to shake things up!

Imagine a Sumo Wrestling Match...

Instead of massive wrestlers, picture tiny atoms jostling for position. Each atom exerts a force, a microscopic push or pull. Sometimes they gang up, pushing together in the same direction. Other times, they clash head-on, their forces battling it out. The resultant force is like the final score of this atomic sumo match – it tells you who's winning and by how much.

Think of it like this:

Force: Imagine a mischievous gremlin trying to move a toy car. He can push it forward, pull it backward, or even sideways. That push or pull is the force, and it has both strength (how hard the gremlin pushes) and direction (which way he's trying to move the car).
Vectors: These are like arrows pointing the way. A long arrow means a strong force, a short arrow means a weak force. And the direction of the arrow? That's where the force is headed!
Scalars: These are just numbers, like the number of cookies in a jar. They don't care about direction, just how much there is.

Calculating the Resultant Force:

Pick a Side: Imagine our toy car is on a tightrope. We can say "right" is positive. Any gremlin pushing to the right gets a plus sign (+), and any gremlin pushing to the left gets a minus sign (-).
Add them Up: Count up all the pushes, remembering the pluses and minuses. That's your resultant force!
Who Wins?
If the answer is positive, the car moves right.
If it's negative, the car goes left.
If it's zero, the gremlins are in a stalemate – the car doesn't budge!

Newton's Second Law: The Cosmic Speed Limit

Sir Isaac Newton, a pretty smart dude, figured out that forces and motion are linked. He said:

$F = ma$

This simple equation is like the cosmic speed limit for the universe. It means:

More Force, More Speed: If our gremlins push the toy car harder (more force), it'll go faster (more acceleration).
Heavy is Slow: If the car is loaded with bricks (more mass), it'll be harder to get it moving, even if the gremlins push with the same force.

This law is everywhere!

Cars: Car designers use it to figure out how strong the engine needs to be to zoom forward or how powerful the brakes should be to screech to a halt.
Rockets: Rocket scientists use it to calculate how much thrust is needed to blast off into space and avoid crashing back down to Earth.
Sports: Coaches use it to analyze how athletes throw a ball, kick a goal, or swing a bat.
Doctors: Doctors use it to understand how our bodies move and how to help us recover from injuries.

Think of it like this:

Imagine you're trying to push a shopping cart. The heavier the cart (more mass), the harder you have to push (more force) to get it moving. And if you want it to go faster (more acceleration), you got to push even harder!

Let's Get Creative!

Airbags: These are like giant pillows that pop out in a car crash. They increase the time it takes for your body to stop, reducing the force and saving you from getting squished.
Rockets: These are like giant firecrackers that blast fuel out the back to create thrust. As the fuel burns, the rocket gets lighter, so it accelerates faster and faster.

Remember:

Units matter! Make sure you're using the right units (Newtons for force, kilograms for mass, and meters per second squared for acceleration).
Draw it out! Sketching a diagram of the forces can help you visualize what's going on.
Friction is a party pooper! In the real world, friction always tries to slow things down.
It's all about the net force! The total force is what matters, not just individual pushes and pulls.
So, there you have it – resultant force and Newton's Second Law, explained with a touch of whimsy and creativity! Hopefully, this makes these concepts a little easier to grasp and a lot more fun to learn about.

Newton's Third Law: It's Not Just About Rockets and Recoil

Okay, let's ditch the textbook jargon and dive headfirst into the wild world of Newton's Third Law! Forget the stuffy "equal and opposite reaction" bit – think of it more like a cosmic dance-off where every move has a consequence.

Imagine this: You're at a party, and you bump into someone. Ouch! But guess what? You stumble backward too. That's Newton's Third Law in action. You pushed them, and they pushed you right back with the same force. It's like the universe is playing a game of invisible tug-of-war, and everything is connected.

Think of it like this:

The Sneeze: You sneeze, and your head jerks back. Why? Because when you sneezed, you actually pushed yourself backward! It's like your own body is giving you a tiny recoil.
The High Five: You and your friend high-five. Whap! You both feel the force of that connection. You pushed their hand, and they pushed yours back. It's a mini explosion of energy!
The Trampoline: You jump on a trampoline, and it throws you back up in the air. It's like the trampoline is saying, "Oh no you don't!" and pushes you away with the same force you pushed on it.

But wait, there's more!

Newton's Third Law isn't just about everyday bumps and jumps. It's the secret behind some of the coolest things in the universe:

Rockets blasting off into space: They push fiery gas down, and the gas pushes them right back up – it's a fiery high five with the cosmos!
Birds soaring through the sky: They push air down with their wings, and the air pushes them right back up – it's like an invisible trampoline in the sky!
Even you, just sitting there: You're pushing down on your chair, and your chair is pushing right back up on you. It's a constant battle against gravity, and you're winning!
So next time you're out and about, remember Newton's Third Law: Every action has a reaction, every push has a push back. It's a cosmic dance, and we're all part of it!

Friction: The Unsung Hero and Villain of Our World

Imagine a world without friction. You'd be sliding all over the place, unable to walk, drive, or even hold a pen! Friction, the force that resists motion between surfaces, is often seen as a nuisance, but it's actually a crucial player in our daily lives.

It's the reason your car stops when you hit the brakes, why you can climb a ladder without slipping, and even why your shoelaces stay tied.

A Microscopic Battleground

Zoom in close enough, and even the smoothest surfaces resemble rugged mountain ranges. When these microscopic peaks and valleys collide, they interlock, creating resistance. The harder you push the surfaces together, the more they grip, and the stronger the friction becomes.

Think of it like trying to walk across a bouncy castle versus a concrete sidewalk. The bouncy castle's surface gives way, making it harder to gain traction. Similarly, rough surfaces create more friction than smooth ones.

Friction's Many Faces

Friction isn't a one-size-fits-all force. It comes in various forms, each with its own quirks:

Static Friction: The stubborn bouncer at the door, preventing an object from moving at all. It's the strongest type of friction and takes the most effort to overcome.
Kinetic Friction: Once you're past the bouncer, kinetic friction is the gentle resistance you feel as you move. It's weaker than static friction, but still noticeable.
Rolling Friction: Imagine pushing a boulder versus rolling it. Rolling friction is like the wheels that make movement smoother and easier.
Fluid Friction: Whether you're swimming in water or flying through the air, fluid friction (also known as drag) is the resistance you feel as you move through a liquid or gas.

The Two Sides of Friction

Friction is a double-edged sword. While it can be a helpful ally, it can also be a frustrating foe.

The Good:

Grip: Friction gives us the grip we need to walk, run, and climb.
Control: Friction allows us to control our movements, whether we're driving a car or writing a letter.
Warmth: Rub your hands together on a cold day, and you'll feel the warmth generated by friction.

The Bad:

Wear and Tear: Friction causes wear and tear on machinery and other objects, leading to damage and reduced lifespan.
Energy Loss: Friction converts kinetic energy into heat, which can be wasteful and even dangerous in some cases.

Friction in Action

Brakes: When you hit the brakes, friction between the brake pads and rotors converts your car's kinetic energy into heat, slowing it down.
Walking: Each step you take relies on friction between your shoes and the ground to propel you forward.
Tires: The friction between your tires and the road allows your car to accelerate, turn, and stop.

Motion with and without Drag

Imagine pushing a box across a frictionless surface. It would continue moving at a constant speed forever! But in the real world, forces like friction and drag always act to slow things down.

Drag is the resistance an object experiences as it moves through a fluid like air or water. The faster you move, the stronger the drag force becomes. Think of a skydiver: as they fall, they accelerate until the drag force equals the force of gravity, reaching a constant speed called terminal velocity.

Friction and Drag in the Real World

Aerodynamic Cars: Car designers use sleek shapes and smooth surfaces to reduce drag and improve fuel efficiency.
Bird Flight: Birds have evolved streamlined bodies and wings to minimize drag and soar through the air with ease.
Projectiles: The trajectory of a bullet or cannonball is affected by both gravity and air resistance.

In Conclusion

Friction and drag are fundamental forces that shape our world. They may be invisible, but their effects are felt everywhere, from the way we walk to the design of airplanes and cars. By understanding these forces, we can appreciate the delicate balance that keeps our world in motion.

1. Reaching Terminal Velocity: A Love Story Between Gravity and Drag

Imagine yourself as a tiny raindrop, leaping from the edge of a cloud. At first, it's pure exhilaration! You're plummeting earthward, the wind whipping past, and the ground rushing up to meet you. This is the thrill of gravity, that invisible force pulling you ever downwards.

But wait! As you pick up speed, you feel a resistance, a gentle pushback against your wild descent. This is drag, like the air itself is trying to hold you aloft. It's a bit like falling in love, really. Gravity is the passionate pull, while drag is the voice of reason, saying "Whoa there, slow down!"

As you fall faster and faster, this push and pull intensifies. It's a cosmic dance, a passionate tango between gravity and drag. Finally, you reach a point of perfect balance, a harmonious equilibrium where the two forces hold each other in a breathless embrace. This, my friend, is terminal velocity – a state of freefall where you're no longer accelerating, but gliding through the air at a constant speed.

But this delicate balance is a fickle thing. It shifts and changes depending on your shape (are you a sleek, aerodynamic raindrop or a fluffy snowflake?), the density of the air around you (think of diving into water versus falling through air), and even your own mass (a feather floats gently, while a rock plummets).

Think of Felix Baumgartner, the daredevil who leaped from the edge of space! Up there, the air is thin, almost non-existent, so gravity had its way with him, pushing him to supersonic speeds. But as he plunged towards Earth, the air thickened, drag increased its grip, and he slowed to a more leisurely pace.

Terminal velocity is everywhere, a hidden force shaping the world around us. It's in the gentle drift of a feather, the graceful arc of a skydiver, and even the pitter-patter of raindrops on your window. It's a reminder that even in the midst of chaos, there's a beautiful balance to be found.

2. Thinking, Braking, Stopping: The Dance of Mind, Machine, and Momentum

Driving is a bit like a three-way conversation between your brain, your car, and the road itself. And when it comes to stopping, this conversation becomes a delicate dance, a symphony of reaction, friction, and physics.

First, there's the "thinking distance," the mental tango. You spot a squirrel darting across the road, and your brain kicks into gear. "Whoa!" it shouts, "Danger ahead!" But like a sleepy dancer, it takes a moment to process the situation and send the signal to your foot to hit the brakes. This delay, this mental cha-cha, is your reaction time. And of course, the faster you're going, the longer this mental dance takes, and the farther you travel before even starting to slow down.

Next comes the "braking distance," where your car takes center stage. The brakes clamp down, the tires grip the road, and friction works its magic, converting your speed into heat and bringing you to a halt. But this dance is influenced by a whole host of factors. Are your brakes sharp and responsive? Are your tires gripping the road like a ballet dancer on pointe shoes, or are they slipping and sliding like someone trying to waltz on ice? And let's not forget the weight of your vehicle – a heavy truck will take much longer to stop than a nimble sports car.

Finally, we have the "stopping distance," the grand finale of this automotive ballet. It's the sum of the thinking distance and the braking distance, the total distance traveled from "Oh no!" to "Whew, that was close!"

Think of accident investigators, like detectives piecing together clues. They measure skid marks, analyze road conditions, and even consider the type of vehicle to reconstruct the events leading up to a crash. They're like choreographers, studying the dance of mind, machine, and momentum to understand how it all went wrong.

Or consider self-driving cars, the futuristic vehicles that promise to revolutionize transportation. These robotic chauffeurs have lightning-fast reflexes, eliminating the "thinking distance" altogether. They're like perfectly synchronized dancers, reacting instantly to every twist and turn of the road.

Ultimately, understanding the dance of thinking, braking, and stopping is about more than just passing a driving test. It's about staying safe, being aware, and respecting the delicate balance between our minds, our machines, and the laws of physics.

Ever slammed on the brakes and wondered how far you actually traveled before coming to a complete stop? That's your stopping distance, my friend, and it's a bit more complex than just how hard you stomp on that pedal. It's like a two-part dance: first your brain has to figure out what's going on (thinking distance), and then your car has to actually wrestle itself to a halt (braking distance).

Think of it like this: you're cruising down the road, jamming to your favorite tunes, when suddenly – BAM! – a rogue squirrel decides to audition for the next Mission: Impossible movie by darting across the road. Your brain goes, "Hold up! Squirrel situation!" That's your thinking distance – the time it takes for your brain to process the fluffy danger and tell your foot to hit the brakes.

But the story doesn't end there. Once your foot slams the brake pedal, your car still needs some space to shake off all that speed and come to a graceful stop. That's your braking distance, and it's affected by a whole bunch of things, from how fast you were going (spoiler alert: faster = longer) to whether the road feels like an ice rink or a cozy carpet.

Let's break it down, shall we?

1. Speed: The Need for Speed... Control, That Is
Speed is the kingpin of stopping distances. Think of it like this: the faster you go, the longer it takes to convince your car to chill out.

Thinking Distance: Imagine you're driving at a leisurely 30 mph, and that squirrel makes its daring dash. Your brain, being the efficient machine it is, takes about a second to react. In that one second, you've already traveled about 13 meters – that's like the length of a bowling lane! Now crank up the speed to 60 mph, and suddenly you're covering two bowling lanes before you even think about braking. Yikes!
Braking Distance: Here's where things get a little more physics-y. The faster you go, the more energy your car has, and that energy needs to go somewhere when you hit the brakes. Doubling your speed doesn't just double your braking distance, it quadruples it! So, if you need 14 meters to stop at 30 mph, get ready for a whopping 56-meter stopping distance at 60 mph. That's like the length of an Olympic-sized swimming pool!
Real-life drama: Remember the tragic 2001 Selby rail crash in the UK? A tired driver accidentally drove onto the train tracks, and a high-speed train couldn't stop in time. Sadly, this highlights how speed can turn a simple mistake into a devastating tragedy.

2. Driver Condition: Are You Fit to Drive?
Your physical and mental state play a huge role in how quickly you react to that surprise squirrel (or any other hazard, for that matter).

Tiredness: Driving drowsy is like driving tipsy – your brain just isn't firing on all cylinders. You're less alert, your reactions are slower, and you might even miss that crucial squirrel altogether. In fact, driving after being awake for 17 hours is like having a few too many drinks!
Alcohol and Drugs: This one's a no-brainer. Alcohol and drugs mess with your judgment, slow down your reactions, and make it harder to control your car. Even a small amount can make a big difference.

Real-life drama: In 2019, a truck driver in the US who hadn't slept in over a day caused a horrific accident that killed seven motorcyclists. A tragic reminder that driving tired can have devastating consequences.

3. Vehicle Condition: Is Your Car Up to the Task?
Your car's health, especially its brakes and tires, is crucial for stopping safely.

Brakes: Think of your brakes as your car's superheroes, swooping in to save the day when you need to stop. But even superheroes need a little TLC. Worn-out brake pads, leaky brake fluid, or any other brake issues can seriously hamper their stopping power.
Tires: Your tires are your car's connection to the road. Worn-out tires are like wearing slippery shoes on a wet floor – you're not going to have much control. And don't forget about tire pressure! Underinflated or overinflated tires can also affect your braking distance.
Load: The more weight your car carries, the harder it is to stop. Think of it like trying to stop a runaway shopping cart versus a tiny toy car.
Real-life drama: In 2018, a tour bus in Mexico crashed, killing 12 people. Faulty brakes were a major factor in the tragedy, highlighting the importance of keeping your car in tip-top shape.

4. Road Conditions: Is it Raining, Shining, or Icy?
The road surface itself can be your best friend or your worst enemy when it comes to stopping.

Wet Roads: Rain creates a slippery layer between your tires and the road, making it harder to grip and stop.
Icy Roads: Ice is like the ultimate braking challenge. It can increase your braking distance by a whopping ten times!
Gravel or Loose Surfaces: Loose gravel or sand can make your tires skid, making it harder to control your car and stop quickly.
Real-life drama: The 2008 M1 motorway crash in the UK involved 34 vehicles and sadly resulted in seven deaths. Fog and icy roads were major culprits, showing just how dangerous bad weather can be.

5. The Wild Cards
A few other things can throw a wrench in your stopping game:

Visibility: Can't see clearly? Whether it's fog, heavy rain, or nighttime driving, poor visibility makes it harder to spot hazards and react in time.
Distractions: Put down that phone! Distracted driving is a major cause of accidents. Anything that takes your attention away from the road increases your reaction time and puts you at risk.
Road Gradient: Going downhill? Gravity's going to lend a helping hand (or rather, a hindering foot) to your braking distance. Going uphill? You'll have an easier time stopping.
The Bottom Line:

Stopping safely is a team effort between your brain, your car, and the road. By understanding these factors and driving responsibly, you can help keep yourself and others safe on the road. So, stay alert, keep your car in good condition, and always be prepared for the unexpected – even if it's just a squirrel with a death wish.

Elastic deformation

Imagine the World as a Giant Playground of Forces

Forget stuffy textbooks and equations for a moment. Think of the world around you – a bustling playground where everything is pushing, pulling, and playfully interacting. That's the magic of forces! Every time you squish a stress ball, bounce a basketball, or even just exist on this planet, you're experiencing the dance of forces.

Shape-Shifting Objects: The Elastic Tango and the Plastic Freeze

Objects are surprisingly flexible, like dancers responding to the rhythm of forces. Sometimes they perform an elegant, reversible tango – that's elastic deformation. Stretch a rubber band, and watch it snap back to its original pose! But sometimes, the music changes, and objects get stuck in a permanent pose – that's plastic deformation. Think of crumpling a piece of paper into a masterpiece (or a mistake!).

What Makes Objects Dance Differently?

Just like dancers have different styles, objects react to forces in unique ways. A feather floats gently in the breeze, while a rock stubbornly resists the wind's push. Why?

The Forceful DJ: The stronger the force, the wilder the dance moves!
Dance Floor Directions: Is the force pushing, pulling, twisting, or squeezing? Each direction creates a different dance move.
Material Moods: A bouncy rubber ball and a rigid steel beam have very different personalities when it comes to forces.
Shape and Size Swagger: A long, thin noodle will bend and sway differently than a chunky block of cheese.

Springs: The Energetic Coils of Fun

Springs are like tiny bundles of coiled-up energy, always ready to bounce back into action. They have a secret code called the spring constant (k) that tells us how much force it takes to stretch or squeeze them. Think of it as a measure of their "springiness"!

Hooke's Law: The Spring's Dance Partner

This law is like a simple dance step for springs: the more you stretch them, the harder they pull back. It's a beautiful balance, but remember, even springs have their limits. Stretch them too far, and they might lose their bounce forever!

Springs in Action: From Bouncy Castles to Tiny Machines

Springs are everywhere! They're the bouncy heart of trampolines and pogo sticks, the shock absorbers in your car, and even the tiny clickers in your pen. Understanding how springs work helps us design amazing things, from comfy mattresses to life-saving medical devices.

Beyond the Basics: Springs Get Serious

Ready to go deeper down the rabbit hole? Explore how springs behave in extreme conditions, how they store energy like tiny batteries, and how they're used in cutting-edge technologies like atomic force microscopes – where they help us see things smaller than we ever imagined!

Keep Exploring!

This is just the beginning of your adventure with forces and springs. There's a whole universe of fascinating concepts waiting to be discovered. So, keep asking questions, keep experimenting, and keep playing in the giant playground of forces!

Imagine a Superhero Stretching Competition

Imagine a superhero stretching competition. Superheroes with amazing elastic powers line up to see who can stretch the furthest. We've got girl, Mr. Fantastic, Plastic Man, and even Stretch Armstrong (he's a superhero in his own right, okay?).

Now, imagine a giant screen displaying a graph for each superhero as they stretch. This is their load-extension graph, a visual representation of how much force (load) it takes to stretch them a certain distance (extension).

The Starting Line: Hooke's Law

At first, everyone stretches in a perfectly straight line. This is the linear elastic region, where the more force they use, the further they stretch in direct proportion. Think of it like a perfect rubber band – double the force, double the stretch. This is where Hooke's Law rules, a simple equation that predicts exactly how far they'll stretch.

Things Get Weird: Non-Linear Elasticity

As the competition heats up, some heroes start to struggle. Girl, for example, finds that it takes more and more effort to stretch further. Her line on the graph starts to curve. This is the non-linear elastic region. She's still stretching, but it's getting harder. The good news is, she can still return to her normal shape when she relaxes.

Pushing the Limits: Plasticity and Fracture

Mr. Fantastic, on the other hand, is feeling confident. He keeps stretching and stretching, his line going way beyond girl. But suddenly, SNAP! He's overdone it and reached his fracture point. He's out of the competition (and might need some superhero first aid).

Plastic Man, being, well, plastic, is showing off. He stretches and stretches, way past where anyone thought possible. His graph enters the plastic region. He's deformed permanently, but hey, he can mold himself back into shape anyway.

The Limit of Proportionality: A Superhero's Limit

The key moment in this competition is the limit of proportionality. This is the point where each hero's graph stops being a straight line. It's the point where Hooke's Law no longer applies, and things get unpredictable.

Why This Matters (Even for Non-Superheroes)

Load-extension graphs aren't just for superhero competitions. They're crucial for understanding how materials behave under stress. Engineers use them to design bridges, buildings, and even those amazing stretchy pants you love.

By understanding the limit of proportionality, engineers can figure out how much force a material can handle before it starts to deform permanently. This is essential for making sure things don't break unexpectedly.

So, next time you see a rubber band, a spring, or even a superhero stretching, remember the load-extension graph. It's a window into the fascinating world of elasticity and the limits of materials.

Circular motion

Imagine a Tiny Dancer Whirling on a Carousel

Picture this: a tiny dancer gracefully twirling on a brightly colored carousel. As she spins, her outstretched arms trace a perfect circle in the air. This captivating scene perfectly illustrates the magic of circular motion, a fundamental dance of physics that plays out on scales both grand and small, from the swirling galaxies in the cosmos to the spinning electrons within an atom.

The Invisible Hand that Guides the Dance

What keeps our tiny dancer moving in her circular path? It's an invisible force, a gentle but firm hand guiding her every twirl. This force, known as centripetal force, always points towards the center of the carousel, like an invisible string connecting the dancer to the very heart of the ride.

A Playful Tug-of-War

Think of circular motion as a playful tug-of-war between two opposing tendencies: the object's natural desire to move in a straight line (like a playful puppy tugging on its leash) and the centripetal force constantly pulling it back towards the center. This delicate balance creates the mesmerizing circular dance.

Let's Explore the Dance Floor

Now, let's explore how different factors influence this captivating dance, without getting bogged down by complicated equations. Imagine we can magically control the force, the dancer's weight, and the size of the carousel.

(a) A Faster Whirl with a Stronger Pull

If we gently increase the pull on our tiny dancer (the centripetal force), she'll start whirling faster, her colorful dress swirling like a vibrant flower. It's like giving her a playful push to make her spin more quickly.

Real-life magic: Think of a race car zooming around a curved track. The friction between the tires and the road acts as the centripetal force, keeping the car from veering off course. The faster the car goes, the stronger the grip needed to stay on track.

(b) A Smaller Circle with a Firmer Grip

If we hold on to our dancer a bit tighter (increasing the centripetal force) but ask her to keep spinning at the same speed, she'll naturally move in a smaller circle, closer to the center of the carousel. It's like drawing her closer for a warm embrace.

Real-life magic: Imagine a satellite orbiting the Earth. The Earth's gravity acts as the centripetal force, keeping the satellite from drifting off into space. If we magically increased Earth's gravity, the satellite would be pulled into a closer, tighter orbit.

(c) A Heavier Dancer Needs a Stronger Hold

If our dancer suddenly became heavier (increased mass) but wanted to keep spinning at the same speed and distance, we'd need to hold on to her more firmly (increase the centripetal force) to keep her from flying off the carousel. It's like holding on tighter to a playful puppy that's suddenly grown bigger.

Real-life magic: Picture two balls of different weights attached to strings and swung in circles. The heavier ball will need a stronger string (more force) to keep it from breaking free and flying away.

The Dance of the Cosmos and Beyond

Circular motion is a universal dance, choreographing the movements of planets around stars, electrons around atoms, and even the swirling patterns of weather systems on Earth. It's a captivating spectacle that reminds us of the intricate and beautiful laws governing our universe.

From Roller Coasters to Washing Machines

This mesmerizing dance isn't just a spectacle of nature; we've harnessed its power in countless inventions. Roller coasters thrill us with their loops and curves, centrifuges help us separate substances in laboratories, and even the spinning drum of a washing machine relies on the principles of circular motion to clean our clothes.

A Never-Ending Waltz

Circular motion is a testament to the delicate balance of forces in our universe, a constant interplay of push and pull that creates a mesmerizing dance of objects both grand and small. It's a dance that has captivated scientists and artists alike for centuries, and its secrets continue to inspire new discoveries and innovations.

Turning effect of forces

Forget Physics Class! The "Moment of Force" Explained with Donuts, Door Slams, and a Tiny T-Rex
Ever felt like physics was just a bunch of equations and boring definitions? Let's ditch the textbook and dive into the "moment of force" with a sprinkle of fun and a dash of the absurd.

1. The "Oh-No-You-Didn't!" Door Slam

Imagine this: you're holding a delicious donut, and your little brother is being extra annoying. You need to shut your bedroom door, fast. Do you gently push the door near the hinges? Nope! You slam it shut by pushing the edge farthest from those hinges. Why? Because that's where your push has the most "oomph" to make the door rotate. That "oomph" is the moment of a force, aka torque.

Think of it like this: the moment of a force is how much a force makes something want to spin. It's like a force's superpower to make things twirl, turn, and rotate.

The Donut-Door Slam Connection:

Force: Your angry push on the door (fueled by donut protection).
Pivot Point: The hinges, where the door spins.

Lever Arm: The distance from your hand to the hinges. The longer the lever arm, the more dramatic the door slam!

2. The T-Rex and the Wrench

Picture a tiny T-Rex trying to loosen a giant nut on a dinosaur-sized bolt. His little arms struggle to turn the wrench. Poor Rex! But wait! He grabs a longer wrench (because even dinosaurs have toolboxes). Suddenly, he's loosening that nut like a pro!

Why? Because the longer wrench gives him a bigger lever arm, increasing the moment of a force he can apply. It's like magic, but it's just physics!

The Moment Equation - No Math Headaches, Promised!

The basic idea is simple:

Moment = Force x Lever Arm

More force, more spin. Longer lever arm, even more spin!

3. Moments in Action - Beyond Donuts and Dinosaurs

Think about it:

Seesaws: A heavy kid needs to sit closer to the middle to balance with a lighter kid further out. It's all about balancing those moments!
Steering Wheels: Turning a car is easier with a bigger steering wheel because you have a longer lever arm.
Bicycles: Pushing the pedals creates a moment that makes the wheels spin.
Bonus Round - Mind-Blowing Moments

Couples: Imagine two forces, equal in strength, pulling in opposite directions. This creates a pure turning force, like when you use a screwdriver.
Center of Gravity: This is the point where an object's weight seems to be concentrated. Its why wobbly things tip over!

The End... or is it?

The moment of a force is everywhere! From opening jars to launching rockets, it's the secret behind making things spin and stay balanced. So next time you see something rotating, remember the "moment of force" and give a little nod to physics. It's way more exciting than you thought, right?

The Balancing Act of Physics: Moments in Motion

Imagine a child teetering on a seesaw, a tightrope walker gracefully balancing high above the ground, or a majestic crane hoisting a massive container into the air. What do all these scenarios have in common? They all showcase the fascinating principle of moments – a fundamental concept in physics that governs how objects balance and rotate.

Moments: The Unsung Heroes of Equilibrium

Before we dive into the principle itself, let's get acquainted with the key players:

Moment: Think of a moment as the "turning power" of a force. It's like an invisible hand trying to twist or rotate an object around a specific point. The greater the force or the farther it's applied from the pivot point, the stronger this turning power becomes.
Equilibrium: This is the state of perfect balance, where an object is either at rest or moving at a constant speed without any change in its rotation. It's like a cosmic tug-of-war where opposing forces perfectly cancel each other out.

The Principle in Action: A Cosmic Balancing Act

The principle of moments states that for an object to be in rotational equilibrium, the sum of all the clockwise moments must be perfectly counterbalanced by the sum of all the anticlockwise moments. It's like a delicate dance of forces, where each twist and turn are met with an equal and opposite reaction.

Moments in Our Everyday Lives

The principle of moments isn't just confined to physics textbooks; it's woven into the fabric of our daily lives:

Seesaws: A classic example! A heavier child needs to sit closer to the fulcrum to balance a lighter child sitting further away. This ensures that the clockwise and anticlockwise moments created by their weights are equal, keeping the seesaw in equilibrium.
Door Handles: Why are door handles placed far from the hinges? Because this increases the moment arm, making it easier to open the door with less force. It's a subtle application of the principle of moments that we often take for granted.
Bicycles: Balancing on two wheels is no easy feat, but the principle of moments plays a crucial role. By subtly shifting their weight and turning the handlebars, cyclists can adjust their center of gravity and maintain equilibrium.

Unveiling the Magic: A Simple Experiment

You can experience the magic of moments firsthand with a simple experiment:

Gather your tools: A meterstick, a triangular prism (or any sturdy pivot), some weights, and some string.
Set the stage: Balance the meterstick horizontally on the pivot, ensuring its perfectly level.
Add the players: Hang weights of different masses on either side of the pivot using the string.
Find the sweet spot: Adjust the position of the weights until the meterstick returns to its balanced state.
Calculate the moments: Multiply the weight of each object by its distance from the pivot.
Witness the magic: The sum of the clockwise moments should be approximately equal to the sum of the anticlockwise moments, showcasing the principle in action.

Beyond the Basics: Delving Deeper into the World of Moments

The principle of moments is a gateway to a deeper understanding of physics and engineering:

Center of Gravity: This is the point where an object's entire weight seems to be concentrated. Understanding the center of gravity is crucial for analyzing stability and balance.
Torque: This is a rotational force that tends to cause an object to rotate. It's closely related to moments and plays a vital role in understanding the mechanics of engines and other rotating systems.
Stability: Objects can be in stable, unstable, or neutral equilibrium. The principle of moments helps us understand how the distribution of weight and the position of the center of gravity affect an object's stability.

The Enduring Legacy of Moments

From the simple act of balancing a book on your head to the complex design of skyscrapers and bridges, the principle of moments is an indispensable tool for understanding and manipulating the world around us. It's a testament to the power of physics to explain the seemingly mundane and the awe-inspiring alike.

Centre of gravity

Centre of Gravity: Where the Earth Gives You a Hug

Imagine the Earth reaching up and giving your favorite mug a warm embrace. That special point where the Earth's hug is most snug? That's the center of gravity! It's like the Earth is saying, "I've got you right here!"

1. The Balancing Point of Existence

Every object, from a tiny pebble to a towering skyscraper, has a secret spot where all its weight seems to gather. It's like the object's personal "belly button," the point where it feels perfectly balanced, no matter how you twist, turn, or hang it.

One and Only: Just like you only have one belly button, every object has just one center of gravity.
Weight's Favorite Hangout: It's the average location of all the tiny bits of weight within an object.
Earth's Influence: The Earth's pull (gravity) plays a big role in deciding where the center of gravity hangs out.

Centre of Gravity vs. Centre of Mass: A Tiny Difference

Most of the time, we can use "center of gravity" and "center of mass" interchangeably, like two peas in a pod. But there's a teeny-tiny difference:

Centre of Mass: Imagine all the "stuff" an object is made of gathered at one point. That's the center of mass. It doesn't care about gravity.
Centre of Gravity: This is where the object's weight feels most concentrated. Gravity is a key player here.

2. Finding the Centre of Gravity: A Playful Experiment

Let's imagine a flat, thin sheet of metal, like a quirky piece of art. How do we find its center of gravity? Time for some playful exploration!

The Plumb Line Game:

Hang It Up: Poke a small hole near the edge of the sheet and hang it freely.
Dangling Detective: Attach a plumb line (a string with a weight) to the same hole.
Trace the Clue: Let the plumb line settle down, then carefully draw a line along its path on the sheet.
Another Go: Repeat steps 1-3, hanging the sheet from a different hole.
X Marks the Spot: The point where your two lines cross? That's the center of gravity!

Why Does This Work?

When the sheet hangs freely, it wants to be as close to the Earth as possible. So, its center of gravity lines up directly below the hanging point. The plumb line, also a fan of gravity, follows the same path. By hanging the sheet from two different points, we create a treasure map, and the "X" marks the hidden treasure: the center of gravity!

3. The Stability Secret

The center of gravity is like an object's secret weapon when it comes to staying upright.

Low and Steady Wins the Race: Objects with a low center of gravity are like sturdy wrestlers; they're tough to knock over. Think of a wide, flat plate.
Top-Heavy Trouble: Objects with a high center of gravity are like wobbly toddlers; they tip over easily. Think of a tall, skinny vase.
The Balancing Act: For an object to stay balanced, an imaginary line from its center of gravity must fall within its base. It's like the object is constantly checking, "Am I still on solid ground?"

Stability Stars:

The Leaning Tower of Pisa: This famous tower is a bit of a rebel, leaning because its foundation is uneven. But it's still standing because its center of gravity, though a bit off-center, keeps it grounded.
Double-Decker Buses: These tall giants have a high center of gravity, making them a bit like giraffes on wheels. Careful design and weight distribution keep them from tipping over.
Ships at Sea: Ships need to be steady even in rough waters. Their design and cargo placement ensure a low center of gravity, keeping them afloat.

4. Centre of Gravity in Action

The center of gravity isn't just a physics concept; it's a real-world superhero!

Engineering Marvels: Engineers use it to design stable buildings, bridges, and vehicles that can withstand the forces of nature.
Athletic Feats: Athletes rely on it to balance, jump, and perform amazing moves. Gymnasts, cyclists, and martial artists are all masters of their center of gravity.

Robot Helpers: Robots use it to navigate and interact with the world, just like we do.
Space Explorers: Spacecraft need to be stable during launch and flight, and the center of gravity plays a crucial role.
Animated Adventures: Animators use it to create lifelike movements for our favorite cartoon characters.
Healthy Living: Doctors and therapists use it to assess balance and help people with mobility challenges.
The center of gravity is a hidden force that shapes our world, from the stability of a simple chair to the design of a soaring skyscraper. It's a reminder that even the most complex things often rely on simple, fundamental principles.

Momentum

Momentum: The Dance of Mass and Motion

Imagine a tiny hummingbird, a blur of iridescent feathers, darting between blossoms. Now picture a massive freight train, rumbling along with unstoppable force. What do these two vastly different objects have in common? Momentum. It's the magic ingredient that keeps things moving, a measure of an object's "oomph" as it travels through space.

Think of it like this: momentum is the stubbornness of an object to change its motion. The hummingbird, light as a whisper, can change direction with the slightest flick of its wings. But that massive train? It needs a whole lot of effort to slow down or speed up. That's because its momentum is huge, thanks to its sheer size and speed.

Impulse: The Art of Changing the Dance

But what happens when we want to change an object's momentum? That's where impulse comes in. Impulse is like a nudge, a push, or a sudden jolt that alters the course of an object's motion. It's the spark that ignites a change in momentum.

Imagine a potter shaping clay on a spinning wheel. With a gentle, sustained push, they can gradually mold the clay into a graceful vase.

That's impulse in action – a force applied over time, subtly transforming the clay's movement and shape.

Or think of a karate chop, a sudden burst of force delivered in an instant. This quick, powerful impulse can shatter a stack of bricks, dramatically altering their momentum from stillness to explosive motion.

The Impulse-Momentum Tango

Impulse and momentum are inseparable dance partners, forever intertwined in the intricate ballet of physics. The impulse-momentum theorem tells us that any change in momentum is the direct result of an impulse. It's like a cosmic cause-and-effect relationship, where every nudge, every push, leaves its mark on an object's motion.

Real-World Rhapsodies

The beauty of impulse and momentum is that they're not just abstract concepts confined to textbooks. They're woven into the fabric of our everyday lives.

The Egg Drop Challenge: Remember that classic science experiment where you had to protect a fragile egg from a fatal fall? The secret to success lies in understanding impulse. By cushioning the egg's landing with soft, yielding materials, you extend the time of impact, reducing the force and preventing a disastrous splat.
The Rocket's Roar: A rocket launch is a symphony of impulse and momentum. The fiery blast of hot gases downwards creates an immense impulse, propelling the rocket upwards with ever-increasing momentum, defying gravity's pull.
The Athlete's Grace: Whether it's a baseball player hitting a home run, a gymnast executing a perfect somersault, or a figure skater gliding across the ice, athletes intuitively understand how to manipulate impulse and momentum to achieve incredible feats of athleticism.

Beyond the Basics

The world of impulse and momentum extends far beyond these examples. It delves into the mysteries of collisions, where momentum is exchanged between objects like secret handshakes. It explores the wonders of angular momentum, the swirling dance of planets and spinning tops. And it even touches upon the profound concept of conservation of momentum, a principle that governs the interactions of everything in the universe.

In Conclusion

Momentum and impulse are not just dry, dusty formulas. They're the dynamic duo of physics, the driving forces behind the motion that fills our world. By understanding their intricate dance, we unlock a deeper appreciation for the elegance and power of the physical universe, from the smallest hummingbird to the vast expanse of space.

Momentum: A Cosmic Dance of Motion and Force

Imagine the universe as a vast, swirling dance floor. Stars, planets, and even tiny particles are all dancers, twirling and gliding in a never-ending cosmic ballet. But what keeps them moving? What dictates their steps and partnerships? The answer lies in the fascinating interplay of momentum and force.

Momentum: The Inertia of Motion

Think of a bowling ball rumbling down the lane. It has a certain "oomph" to its movement, a resistance to change. That's momentum! It's the measure of an object's motion, a combination of its mass and velocity. The heavier and faster an object is, the more momentum it carries.

Force: The Choreographer of Change

Now, imagine that bowling ball hitting the pins. Suddenly, there's a chaotic flurry of movement! That's force in action. Force is like the choreographer of this cosmic dance, capable of changing an object's momentum – speeding it up, slowing it down, or altering its direction.

The Conservation Principle: A Universal Pact

But there's a rule in this cosmic dance floor: momentum can never be created or destroyed, only transferred. This is the principle of conservation of momentum. It's like a universal pact that ensures the total amount of "oomph" in the universe remains constant.

A Billiards Analogy

Picture two billiard balls colliding. Before the collision, each ball has its own momentum. When they clash, they exchange some of that momentum. One might speed up, the other might slow down, or they might even change direction. But the total momentum of the system? It remains the same.

Newton's Second Law: The Force-Momentum Connection

Sir Isaac Newton, a keen observer of this cosmic dance, discovered a fundamental link between force and momentum. He found that the force acting on an object is directly proportional to the rate at which its momentum changes. In other words, the harder you push something, the faster its momentum changes.

Real-World Applications: From Rockets to Crumple Zones

This interplay of momentum and force isn't just theoretical. It's at play all around us!

Rocket Propulsion: Rockets harness the conservation of momentum. By expelling hot gases downward, they gain an equal and opposite momentum, propelling them upward into the vastness of space.
Car Safety: Crumple zones in cars are designed to increase the time of impact during a collision. This reduces the force experienced by the passengers, protecting them from harm.
Sports: Whether it's a baseball bat hitting a ball or a golfer driving a ball off the tee, the transfer of momentum is key to achieving the desired outcome.

The Dance Continues...

The interplay of momentum and force is a fundamental principle that governs the motion of everything in the universe. It's a captivating dance of mass, velocity, and interaction, a testament to the elegant laws of physics that orchestrate the cosmos.

Energy, work and power

Energy

Energy: The Universe's Hidden Dance

Imagine the universe as a grand ballroom, filled with dancers swirling and twirling in an endless waltz. But instead of music, it's energy that fuels this cosmic dance – an invisible force, a magical current, constantly shifting and changing, yet never disappearing. This is the law of conservation of energy: the fundamental rule that governs this grand cosmic ball.

Where Does the Energy Hide?

Just like dancers need a place to rest between dances, energy needs places to "hang out" – we call these energy stores. Let's peek into some of the most popular spots in this cosmic ballroom:

The Kinetic Dance Floor: This is where the action happens! Any dancer in motion – a speeding comet, a swirling electron, even a toddler on a sugar rush – is showing off their kinetic energy. The faster the dance, the more energy is unleashed! Think of a spinning top, a rushing river, or a rocket blasting off – all brimming with kinetic energy.

The Gravity Balcony: Ever watched a graceful ballet leap? That dancer seems to hang in the air, defying gravity. That's gravitational potential energy! The higher the leap, the greater the potential energy stored. Imagine a book perched on a high shelf, water held back by a mighty dam, or a satellite orbiting the Earth – all holding onto this "high-up" energy.

The Chemical Cocktail Bar: This is where energy gets mixed up and shaken, not stirred! It's the energy hidden within the bonds of molecules, like secret ingredients in a potent cocktail. When these bonds break, the energy explodes outwards, sometimes as heat, sometimes as light. Think of the fiery explosion of fuel igniting, the slow burn of food giving you energy, or the sudden burst of light from a firework – all thanks to chemical energy.

The Elastic Stretch Lounge: Ever stretched a rubber band until it almost snaps? That's elastic energy being stored! It's the energy of things being squished, stretched, or twisted. Imagine a spring coiled tight, a bow pulled taut, or a trampoline bouncing back – all examples of this "springy" energy.

The Nuclear Powerhouse: This is the heart of the universe, the core of all energy! It's the energy locked deep within the heart of atoms, a powerhouse waiting to be unleashed. Nuclear reactions, like those in the sun or a nuclear power plant, release this energy with incredible force. But beware, this energy can be dangerous if not handled carefully, like the tragic reminder from the Fukushima disaster.

The Electrostatic Spark Room: Ever felt a static shock on a dry winter day? That's electrostatic energy – the energy of electric charges playing tag. It's the zap of lightning, the cling of clothes, the crackle of a spark. This energy is used in all sorts of cool gadgets, from photocopiers to touchscreens.

The Thermal Sauna: This is where things get toasty! It's the energy of heat, the vibration of molecules within an object. The hotter something is, the more thermal energy it has. Imagine a steaming cup of coffee, a glowing ember, or the Earth's fiery core – all radiating thermal energy.

Energy on the Move!

But energy isn't just about sitting still in these stores. It loves to travel, to flow from one place to another. Think of it as the dancers' switching partners, moving from one dance to the next. Here are some of the ways energy gets around this cosmic ballroom:

Mechanical Work: The Push and Pull: Imagine pushing a heavy box across the floor, or lifting a weight high above your head. That's mechanical work – energy being transferred by force. Think of a hydroelectric dam harnessing the power of falling water, or a car engine converting fuel into motion.

Electrical Work: The Current Affair: Electricity is like a river of energy, flowing through wires and circuits, powering our world. Think of a light bulb glowing, a motor spinning, or a phone charging – all thanks to the flow of electrical energy.

Heating: The Warm Embrace: Heat is like a cozy hug, transferring energy from a hotter object to a colder one. Imagine a hot stove warming a pan, a radiator heating a room, or the sun's rays warming the Earth.

Waves: The Ripple Effect: Energy loves to travel in waves, like ripples spreading across a pond. Light, sound, even earthquakes – they all carry energy in waves. Think of the warmth of sunlight, the sound of music, or the destructive power of a tsunami.

The Grand Finale

So, as the cosmic dance continues, remember that energy is the driving force, the hidden choreographer. It's in everything we see, hear, and feel. By understanding its different forms and how it moves, we can unlock the secrets of the universe and create a more sustainable future. After all, we're all part of this grand dance, connected by the flow of energy that binds us together.

Energy: The Universe's Dance of Transformation

Imagine the universe as a grand, never-ending dance party. The music pulses, the lights flash, and the dancers twirl and leap with boundless energy. But where does all this energy come from? And where does it go when the music fades?

The answer lies in a fundamental principle of physics: the Principle of Conservation of Energy. This principle tells us that energy, like the dancers at our cosmic party, never truly disappears. It simply changes form, shifting from one dazzling move to another.

Energy's Many Masks

Think of energy as a master of disguise. It can manifest in countless ways:

Kinetic Energy: The vibrant energy of motion, like the whirling of a dancer or the rush of a roller coaster.
Potential Energy: Stored energy, like a coiled spring ready to unleash its power or a boulder perched on a cliff edge.
Thermal Energy: The heat that warms our bodies, fuels our engines, and powers the sun.
Chemical Energy: The energy locked within the bonds of molecules, released when we burn fuel or digest food.
Nuclear Energy: The immense power contained within the heart of atoms, driving the stars and nuclear power plants.
Electrical Energy: The flow of electrons that illuminates our homes and powers our devices.
Radiant Energy: The energy that travels in waves, like the light that bathes our world and the radio waves that carry our music.

The Unbreakable Rule

The Principle of Conservation of Energy states that in this grand cosmic dance, no energy is ever lost or created. It simply transforms from one form to another, like a dancer seamlessly transitioning from a waltz to a tango.

Imagine a child on a swing. As they soar upwards, their kinetic energy (energy of motion) is converted into potential energy (stored energy). At the peak of their arc, they pause for a moment, all kinetic energy transformed into potential energy. Then, they swing back down, potential energy flowing back into kinetic energy.

This principle applies to everything in the universe, from the smallest atom to the largest galaxy. It's a fundamental truth that underpins our understanding of how the world works.

Energy in Our World

The Principle of Conservation of Energy isn't just an abstract concept. It has real-world implications:

Fuel-Efficient Cars: Hybrid and electric cars are designed to minimize energy waste, converting more of the energy stored in fuel into motion.
Green Buildings: Architects use energy-efficient materials and designs to reduce energy consumption and create sustainable living spaces.
Renewable Energy: Solar panels, wind turbines, and hydroelectric dams harness the power of nature to generate clean energy, reducing our reliance on fossil fuels.

The Quest for Perpetual Motion

The Principle of Conservation of Energy also tells us that a perpetual motion machine, a device that can run forever without any energy input, is impossible. Energy cannot be created from nothing; it can only be transformed.

A Universe in Balance

The Principle of Conservation of Energy reveals a universe in constant motion, a place where energy flows and transforms in an intricate dance. It's a reminder that everything is connected, and that energy, like life itself, is a precious resource that must be used wisely.

Kinetic Energy: The Zing of Things in Motion

Imagine the world as a grand, bustling dance floor. Everything that moves, from the tiniest dust mote swirling in a sunbeam to a rocket soaring through the cosmos, is grooving to the rhythm of kinetic energy. It's the energy of motion, the vibrant pulse of the universe!

The Secret Formula: Unmasking Kinetic Energy

Scientists, those ever-curious explorers of the natural world, have devised a clever way to measure this energetic dance. They use a special formula, a kind of secret code that unlocks the mysteries of motion:

KE = 1/2 * m * v^2

Let's break it down, shall we?

KE: This stands for kinetic energy, the star of our show. It's measured in joules (J), a unit named after James Prescott Joule, a brilliant physicist who dedicated his life to understanding energy.
m: This represents mass, the amount of matter in an object. Think of it as the object's "heft" or "substance." The more massive an object, the more kinetic energy it carries when in motion.
v: This is velocity, the speed at which an object moves in a specific direction. It's not just about how fast something goes, but also where it's headed.

Unveiling the Magic: What the Formula Reveals

This seemingly simple formula holds profound secrets about the nature of kinetic energy:

Mass Matters: Kinetic energy is directly proportional to mass. Imagine two dancers, one light and nimble, the other a powerful heavyweight. If they move at the same speed, the heavyweight dancer possesses more kinetic energy.
Velocity Rules: Kinetic energy is even more closely linked to velocity. In fact, it's proportional to the square of the velocity. So, if you double an object's speed, its kinetic energy quadruples! This explains why a speeding car can cause so much damage in a collision.
Magnitude Only: Kinetic energy is a scalar quantity, meaning it only cares about magnitude, not direction. It's like the volume of music on the dance floor; it doesn't matter if the dancers are moving left, right, or spinning in circles, the energy is still there.

Kinetic Energy in Action: A World in Motion

Look around you! Kinetic energy is everywhere:

The graceful arc of a baseball: A pitcher throws a fastball, transferring kinetic energy to the ball. The batter swings, adding even more energy, sending the ball soaring.
The rush of a roller coaster: As the coaster plunges down a steep hill, potential energy transforms into kinetic energy, giving riders an exhilarating thrill.
The gentle breeze rustling leaves: Air molecules, energized by the sun, dance and collide, creating the wind that whispers through the trees.
The bustling city: Cars, buses, and people, all in motion, contribute to the kinetic energy of urban life.

Beyond the Basics: Delving Deeper

Kinetic energy isn't just about objects moving in straight lines. It also encompasses:

Rotational kinetic energy: The whirling of a dancer, the spin of a planet, the twirl of a fidget spinner – all possess rotational kinetic energy.
Vibrational kinetic energy: The quivering of a guitar string, the oscillation of atoms in a solid, the jiggling of molecules in a hot cup of tea – these are all examples of vibrational kinetic energy.

Kinetic Energy and its Transformations: The Dance of Energy

Kinetic energy is a master of disguise, constantly transforming into other forms of energy:

Potential energy: A ball held high in the air has potential energy. As it falls, this potential energy converts into kinetic energy.
Thermal energy: Friction between moving objects generates heat, a form of kinetic energy at the molecular level.
Sound energy: The vibration of a drumhead produces sound waves, carrying kinetic energy through the air.
Electrical energy: The flow of electrons in a wire is a form of kinetic energy that powers our homes and devices.

The Grand Finale: Kinetic Energy's Cosmic Dance

From the smallest subatomic particles to the largest galaxies, kinetic energy is the driving force behind the universe's endless dance. It's a fundamental concept that helps us understand the world around us, from the everyday to the extraordinary. So, next time you see something in motion, take a moment to appreciate the magic of kinetic energy – the zing of things in motion!

Imagine a world where gravity is a mischievous imp, always playing tricks. It loves to tug on things, pulling them closer to the ground. The higher you lift something, the more this imp strains and grumbles, storing up its energy to pull it back down. This stored energy, my friend, is gravitational potential energy.

Think of it like this: every object has a tiny, invisible spring attached to it, connecting it to the Earth. The higher you lift the object, the more you stretch this spring. The imp, of course, is hiding inside the spring, getting angrier and angrier. This "anger" is the potential energy, waiting to be unleashed!

The Equation: $\Delta Ep = mg\Delta h$

Now, let's peek behind the curtain and see how this imp's mischief is measured. Our magical formula is $\Delta Ep = mg\Delta h$. Let's break it down:

ΔEp (Change in Gravitational Potential Energy): This is how much the imp's anger changes when you move something. It's measured in Joules (J), which are like little units of "oomph" the imp has stored up.
m (Mass): Imagine trying to lift a feather versus a bowling ball. The bowling ball is much heavier, right? That's mass, and the more mass an object has, the stronger the imp's pull and the more potential energy it stores.
g (Acceleration due to Gravity): This is the imp's strength! On Earth, it's about $9.8\ m/s^2$, kind of like the imp's muscle power.
Δh (Change in Height): This is how far you stretch the invisible spring. The higher you lift something, the more potential energy the imp stores.

The Imp's Games: Key Concepts

Work-Energy Theorem: Every time you lift something against the imp's pull, you're doing work. This work is stored as potential energy in that stretched spring. It's like a cosmic tug-of-war!
Conservation of Energy: Energy can't be created or destroyed, just transformed. So, the imp's potential energy can be converted into other forms of energy, like the energy of motion (kinetic energy). Think of a ball rolling downhill – the imp's stored energy is released, making the ball go faster and faster!
Zero Point: Where does the imp's influence end? Well, that's up to you! You can choose any point as the "zero" for potential energy. It's like deciding where to start measuring the spring's stretch.

The Imp in Action: Examples

Lifting a Book: When you lift a book, you're stretching the spring and storing energy. The imp is grumbling, waiting to pull it back down.
Roller Coaster: The climb up the first hill is all about storing potential energy. Then, whoosh! The imp takes over, releasing that energy as the coaster races down.
Hydroelectric Dam: Water high up behind a dam has a lot of potential energy. When the water flows down, the imp's energy is released, turning turbines and generating electricity.

The Imp's Tricks: Advanced Concepts

Gravitational Potential: This is like a map of the imp's influence. It tells you how much potential energy an object would have at any point in space.
Escape Velocity: If you throw something fast enough, you can overcome the imp's pull and escape its grasp entirely! That's escape velocity.
Orbits: Planets and satellites are constantly playing a game of tag with the imp. They're moving so fast that they keep falling but never quite hit the ground.

The Imp's Legacy

Gravitational potential energy is everywhere, from the swing of a pendulum to the orbits of planets. It's a fundamental force of nature, and understanding it helps us understand the universe around us. So next time you lift something, remember the mischievous imp, storing up its energy, waiting to play its tricks!

Work

Work: It's Not Just About Elbow Grease!
Imagine a lazy afternoon. You're sprawled on the couch, remote in hand, flipping through channels. Are you doing any work? Physics might surprise you! While you're chilling, your heart is pumping blood, your lungs are expanding and contracting, and tiny motors within your cells are buzzing with activity. That, my friend, is work!

Work, in the physics sense, happens whenever a force gets something moving. Think of it like this: you push a swing (force), and the swing goes higher (movement). You've done work! The harder you push and the higher it swings, the more work you do.

But here's the catch: the force and the movement have to be in the same direction. Pushing a wall with all you might feel like work, but if the wall doesn't budge, you haven't actually done any work according to physics. Sneaky, right?

Let's break it down with some everyday examples:

The Gym Rat: Every time you lift that dumbbell, you're doing work against gravity. The heavier the weight and the higher you lift it, the more work you do. Feel the burn!
The Bookworm: Even lifting a book requires work. You're applying a force to overcome gravity and move the book upwards. So yes, reading can be a workout (for your brain and your biceps!).
The Speedy Cyclist: As you pedal your bike, you apply a force to the pedals, causing the wheels to turn and propel you forward. The faster you go and the further you travel, the more work you do.

Work isn't just about physical exertion, though. Think about a car engine burning fuel to move the car, or a crane lifting heavy loads on a construction site. These machines are doing work too, converting energy into motion.

Even the tiniest things do work. Imagine a little ant carrying a crumb many times its size. That ant is a champion weightlifter in the insect world!

So next time you hear the word "work," don't just think about chores or your job. Remember the amazing ways that forces and motion interact to create the world around us, from the grand scale of planets orbiting stars to the microscopic dance of molecules. It's a universe of work out there!

Energy resources

The Energy Tango: A 2024 Guide to Powering Our World
Forget dry textbooks and yawn-inducing lectures! Let's dive into the electrifying world of energy, where the old guard of fossil fuels faces off against the rising stars of renewables. Think of it as an epic dance-off for the future of our planet.

Renewables: The Upbeat Underdogs

These sources are like nature's party animals – they keep the good times rolling, constantly renewing themselves. They're clean, green, and ready to take center stage.

Solar: Imagine tiny sun-catchers blanketing rooftops, soaking up rays and spitting out electricity. That's solar power in a nutshell! But hold on, it gets even cooler. In 2024, we're talking perovskite solar cells – sleek, efficient, and cheaper than ever. Picture skyscrapers shimmering with built-in solar skins (BIPV) and farms where crops and solar panels share the land in a harmonious Agri voltaic embrace. Dubai's Mohammed bin Rashid Al Maktoum Solar Park is a shining example of Solar's potential to light up entire cities.

Wind: Giant windmills gracefully pirouette in the breeze, transforming gusts into gigawatts. But these aren't your grandpa's windmills. We're talking towering turbines with blades longer than a jumbo jet, floating majestically offshore. They're smarter too, with cutting-edge control systems that optimize energy capture in every gust and swirl. The UK's Hornsea Project 2, a colossal offshore wind farm, shows how wind power can blow away our reliance on fossil fuels.

Hydro: Water, water everywhere, and not a drop to waste! Hydropower harnesses the raw power of moving water. Think massive dams holding back raging rivers, sleek turbines spinning in the current, and ingenious pumped storage systems that act like giant batteries. But it's not just about brute force. In 2024, hydropower is getting a makeover with fish-friendly turbines and eco-conscious flow management. The mighty Itapúa Dam, a testament to hydropower's might, illuminates' homes across Brazil and Paraguay.

Biomass: Turning trash into treasure! Biomass energy converts organic matter – wood, crops, even agricultural leftovers – into fuel. Think high-tech biofuel refineries brewing up green gasoline from algae and biogas plants churning out energy from food waste. Sweden, a biomass champion, shows how to heat homes and power industries with sustainable forestry.

Geothermal: Tapping into Earth's inner furnace! Geothermal power plants harness the heat simmering beneath our feet, turning it into electricity. But the real action is in Enhanced Geothermal Systems (EGS), which unlock energy from hot dry rocks, expanding geothermal reach. Iceland, a geothermal paradise, proves that Earth's heat can keep us warm and powered up.

Non-Renewables: The Fading Rockstars

These guys were the headliners for centuries, but their time in the spotlight is coming to an end. They're dirty, finite, and leaving a trail of damage in their wake.

Coal: The old king of fossil fuels, coal is facing a bleak future. Its reign is marred by pollution, health hazards, and a shrinking fan base.

China, the world's coal kingpin, is feeling the pressure to clean up its act.

Oil: The liquid lifeline of our transportation systems, oil is a volatile character. Price swings, environmental disasters, and dwindling reserves are casting a shadow on its future. The Middle East, holding a lion's share of oil reserves, plays a pivotal role in this ongoing drama.

Natural Gas: The "cleaner" fossil fuel, natural gas still has its share of problems. Methane leaks, fracking controversies, and the rise of renewables are challenging its dominance. The US shale gas boom offers a glimpse into both the potential and pitfalls of natural gas.

Nuclear: A controversial figure, nuclear power boasts zero greenhouse gas emissions but carries the baggage of radioactive waste and the specter of catastrophic accidents. While it still plays a role in the energy mix, high costs and safety concerns are holding it back.

The Encore:

The energy landscape is in constant flux. Renewables are gaining ground, driven by innovation, falling costs, and growing environmental awareness.

While fossil fuels cling to their fading glory, their days are numbered. The future belongs to clean, sustainable energy sources that can power our world without compromising the planet. So, let's get this dance party started and usher in a brighter, greener future!

The Energy Within: From Ancient Sunlight to Electrified Cities

Imagine the Earth millions of years ago, teeming with life. Lush forests, vast swamps, and oceans brimming with tiny creatures. As these organisms lived and died, their remains sank to the depths, slowly transforming under immense pressure and heat. This, my friend, is the origin story of fossil fuels – pockets of ancient sunlight trapped within the Earth, waiting to be unleashed.

Fossil Fuels: A Legacy of Energy

These fuels – coal, petroleum, and natural gas – are like time capsules, holding the energy of bygone eras. We access this energy through a process called combustion, essentially controlled burning.

Coal: Picture massive power plants where coal is fed into fiery furnaces, heating water to create steam that roars through turbines, spinning them like tireless giants. These turbines, in turn, power generators that bring electricity to our homes and cities.
Petroleum: Think of the gasoline that fuels your car or the diesel that powers trucks and ships. These refined forms of petroleum explode within internal combustion engines, creating the mechanical force that drives our modern world.
Natural Gas: This cleaner-burning fossil fuel is used in power plants similar to coal, or piped directly into homes for heating and cooking.
But our relationship with fossil fuels is complex. While they've powered incredible progress, they also leave a mark on our planet.

The Environmental Cost:

Burning fossil fuels releases greenhouse gases, like invisible blankets trapping heat and warming the Earth. This contributes to climate change, with consequences like rising sea levels, extreme weather, and disruptions to delicate ecosystems. It's like we're borrowing energy from the past but leaving a hefty bill for future generations.

The Quest for Cleaner Solutions:

Scientists and engineers are constantly searching for ways to reduce our reliance on fossil fuels and mitigate their impact.

Carbon Capture and Storage (CCS): Imagine capturing the carbon dioxide emissions from power plants and storing them deep underground, preventing them from reaching the atmosphere. It's like giving those emissions a permanent time-out!
Combined Cycle Gas Turbine (CCGT): These power plants are like the hybrid vehicles of the energy world, combining gas and steam turbines to squeeze out more electricity from the same amount of fuel.

Biofuels: Nature's Renewable Energy

But what if we could harness the energy of the present, not just the past? Enter biofuels – fuels derived from recently living organisms like plants and algae.

Ethanol: Think of cornfields swaying in the breeze, or sugarcane plantations bathed in sunlight. These plants can be processed to produce ethanol, a type of alcohol that can be blended with gasoline or used to power special vehicles.
Biodiesel: Imagine using vegetable oils, animal fats, or even recycled cooking grease to fuel trucks and buses! That's the power of biodiesel.
Biogas: This fuel is created from the breakdown of organic matter, like agricultural waste and food scraps. It's like nature's recycling system, turning waste into a valuable energy source.

The Biofuel Balancing Act:

While biofuels offer a greener alternative, they come with their own set of challenges. We need to ensure that growing crops for fuel doesn't compete with food production or lead to deforestation. It's all about finding a sustainable balance.

A Diverse Energy Future:

The future of energy is likely to be a vibrant tapestry, woven from many sources. Fossil fuels may still play a role, but their dominance will likely fade as we embrace cleaner, more sustainable options. Biofuels, solar power, wind energy, and other renewable sources will take center stage, powering a brighter future for generations to come.

The Dance of Water and Sunlight: A Tale of Two Energies

Imagine a world where the very rhythm of nature powers our lives. Where the graceful dance of water and the fiery kiss of the sun become sources of clean, abundant energy. This isn't a fantasy, but a reality unfolding before our eyes with the power of hydroelectric and solar energy.

Hydropower: The River's Symphony

Picture a river, a powerful force of nature, carving its path through mountains and valleys. For centuries, we've harnessed this raw energy, transforming it into a symphony of electricity.

Think of a hydroelectric dam as a conductor, orchestrating the flow of water. The dam creates a reservoir, a serene lake holding the potential energy of millions of water droplets. As these droplets are released, they surge through massive penstocks, like eager musicians rushing to the stage.

Finally, they reach the turbines, the heart of the hydropower plant. These giant wheels spin with incredible force, driven by the water's momentum. Connected to generators, they convert this mechanical dance into a surge of electrical power, a harmonious melody that flows through power lines to light up our homes and cities.

Solar Power: Capturing the Sun's Fire

Now, let's shift our gaze to the sky, where the sun, a celestial artist, paints the world with light and warmth. Solar energy captures this radiant energy, transforming it into a vibrant source of power.

Imagine countless photovoltaic cells, tiny alchemists, basking in the sun's glow. When sunlight strikes these cells, a magical transformation occurs. Photons, particles of light, knock electrons loose, creating a flow of electricity. It's like capturing the sun's fire and channeling it into a steady current.

From rooftops adorned with solar panels to vast solar farms stretching across deserts, we're harnessing the sun's power on a grand scale. This clean, abundant energy source is a beacon of hope, illuminating a future free from the shackles of fossil fuels.

A Duet for a Sustainable Future

Hydropower and solar energy are not competitors, but partners in a duet for a sustainable future. They complement each other, offering a diverse and resilient energy mix.

Hydropower provides a steady, reliable baseload power, while solar energy shines during peak demand hours. Together, they create a harmonious balance, ensuring a clean and reliable energy supply for generations to come.

Beyond the Basics: Innovations and Inspiration

The world of hydropower and solar energy is constantly evolving. Scientists and engineers are pushing the boundaries of innovation, developing:

Fish-friendly turbines: Allowing aquatic life to navigate safely through hydroelectric dams.
Variable speed turbines: Optimizing energy production based on river flow.
Higher efficiency solar cells: Capturing more sunlight and generating more electricity.
Building-integrated photovoltaics: Seamlessly integrating solar panels into building designs.
These advancements are not just technological marvels; they are testaments to human ingenuity and our commitment to a sustainable future.

Join the Movement

The journey towards a clean energy future is a collective one. Embrace the power of water and sunlight. Support renewable energy initiatives. Together, we can create a world where the rhythm of nature powers our lives, leaving a legacy of sustainability for generations to come.

Powering Our World: A Tale of Two Earths

Imagine our planet as a giant, steaming cauldron. Deep within its fiery heart, immense forces churn, remnants of its explosive birth. This primal energy, the very essence of Earth's being, offers us two incredible gifts: the atom's might and the Earth's own inner fire.

1. Nuclear Fuel: Taming the Tiny Titans

Picture the atom, an infinitesimal speck of existence, yet holding within it the power of a thousand suns. We've learned to unlock this power, to coax the atom into releasing its captive energy through a process called nuclear fission. It's like splitting a tiny, incredibly dense seed, unleashing a torrent of energy that can light up entire cities.

The Dance of the Neutrons: Imagine a microscopic billiards game, where neutrons, tiny subatomic particles, smash into atoms, splitting them apart and releasing more neutrons in a chain reaction. This controlled atomic dance, within the heart of a nuclear reactor, generates immense heat.

From Heat to Light: This heat transforms water into steam, a powerful force that spins turbines connected to generators, ultimately producing the electricity that powers our homes, schools, and hospitals. It's a marvel of engineering, a testament to our ability to harness the fundamental forces of the universe.

But this power comes with a heavy burden:

The Ghost of Accidents Past: The names Chernobyl and Fukushima echo with the chilling reminder of what can happen when this atomic dance spirals out of control.

The Legacy of Waste: The spent nuclear fuel, still radioactive, remains a challenge, demanding secure storage solutions for generations to come.

Yet, the atom's allure remains strong:

New Age Reactors: Scientists are forging ahead, designing smaller, safer, and more efficient reactors. Imagine modular reactors, like atomic Lego blocks, that can be tailored to specific energy needs.

The Dream of Fusion: The holy grail of energy, nuclear fusion, mimics the very process that powers the sun. It promises clean, abundant energy, but its realization remains a tantalizing challenge for future generations.

2. Geothermal Energy: Earth's Warm Embrace

Beneath our feet, the Earth's core burns with a fiery intensity. This heat, radiating outwards, creates pockets of scalding water and steam trapped within the Earth's crust. Imagine these as subterranean cauldrons, brimming with natural energy, waiting to be tapped.

Tapping the Earth's Blood: We drill deep into these reservoirs, extracting the hot water or steam to drive turbines and generate electricity. It's like harnessing the Earth's own heartbeat to power our world.
More Than Just Electricity: This geothermal energy can also directly heat our homes, greenhouses, and even provide soothing warmth to natural hot springs.

Geothermal energy offers a gentler touch:

A Constant Source: Unlike the fickle sun or wind, the Earth's heat is always there, a reliable source of power, day and night.
Harmony with Nature: Geothermal power plants have a minimal environmental impact, releasing far fewer greenhouse gases than fossil fuels.

But challenges remain:

Location, Location, Location: Geothermal resources are not evenly distributed, limiting their accessibility.
Earth's Grumbles: Extracting geothermal energy can sometimes disturb the Earth's delicate balance, triggering minor tremors.

The future of geothermal is bright:

Enhanced Systems: Scientists are developing ways to tap into even deeper, hotter rocks, expanding the potential of geothermal energy.
Heating Our Homes: Geothermal heat pumps are gaining popularity, offering an energy-efficient way to keep our homes cozy in winter and cool in summer.

A Symbiotic Future:

Both nuclear and geothermal energy offer powerful pathways to a cleaner, more sustainable future. They are testaments to human ingenuity, our ability to harness the Earth's resources and the universe's secrets to power our world. As we continue to innovate and explore, these two powerful forces will undoubtedly play a crucial role in shaping our energy landscape for generations to come.

Nature's Symphony: How Wind, Tides, and Waves are Powering Our World

Imagine a world where the very forces that shape our planet also power our homes and cities. That's the promise of renewable energy sources like wind, tides, and waves – a symphony of nature orchestrated to provide clean, sustainable power.

Wind Energy: Dancing with the Breeze

Picture a field of graceful giants, their blades pirouetting in the wind like dancers. These are wind turbines, harnessing the invisible force of moving air. The sun's uneven embrace of the Earth creates this ballet of air pressure, and we, with our ingenious turbines, capture its energy.

The Wind Turbine's Inner Workings:

Rotor: Think of it as the dancer's arms, catching the wind and transforming it into a swirling motion.
Gearbox: Like a conductor speeding up the tempo, it increases the rotor's spin to match the generator's rhythm.
Generator: The heart of the operation, converting the dance of motion into electrical energy.
Nacelle: The turbine's head, housing the conductor (gearbox) and the heart (generator).
Tower: A sturdy leg, lifting the dancer high to catch the strongest breezes.

Types of Wind Turbines:

HAWTs: The classic ballerina, spinning gracefully on a horizontal axis.
VAWTs: A more modern dancer, twirling with vertical flair.

Wind Energy's Evolution:

Giants in the Sky: Modern turbines have grown into towering titans, capturing more of the wind's power.
Ocean's Embrace: Offshore wind farms venture into the sea, where the wind is a powerful and constant partner.
Floating Wonders: Like lilies on a pond, floating turbines dance on the water's surface, expanding our reach.
Smart Connections: Wind energy joins the "smart grid," a network that intelligently distributes power where it's needed most.

Wind's Success Stories:

Hornsea Two (UK): A majestic wind farm in the North Sea, a testament to human ingenuity.
Gansu Wind Farm (China): A vast project harnessing the winds of the Gobi Desert, a symbol of China's commitment to renewable energy.

Tidal Energy: Riding the Ocean's Rhythm

The moon, a celestial conductor, orchestrates the ocean's tides. Twice a day, it pulls the sea in a rhythmic dance of rise and fall. We, with our tidal power technologies, capture this predictable energy.

Tidal Power Trio:

Tidal Barrages: Like dams, they capture the high tide and release its energy through turbines as the tide recedes.
Tidal Stream Generators: Underwater turbines that spin like underwater windmills, driven by the ocean's currents.
Tidal Lagoons: Artificial coves that capture the tide's energy, like miniature, controlled seas.

Tidal Energy's Strengths and Challenges:

Strengths: Predictable, powerful, and with minimal impact on the environment.
Challenges: High initial costs and the need for specific geographical conditions.

Tidal Pioneers:

La Rance (France): The world's first tidal power plant, a testament to human innovation.
Sahwa Lake (South Korea): The world's largest tidal power station, harnessing the power of the Yellow Sea.

Wave Energy: Tapping the Ocean's Pulse

The ocean's surface is a restless canvas, painted with waves generated by the wind. Wave energy converters are like skilled artisans, capturing the energy of these rolling hills and valleys.

Wave Energy's Diverse Toolkit:

Point Absorbers: Buoys that bob and weave with the waves, driving generators within.
Oscillating Water Columns: Chambers that capture the waves' breath, compressing air to drive turbines.
Overtopping Devices: Structures that channel waves into reservoirs, releasing their energy through turbines.
Attenuators: Long, segmented structures that flex with the waves, generating power from their motion.

Wave Energy's Promise and Hurdles:

Promise: Abundant, widely available, and environmentally friendly.
Hurdles: Technological challenges and high costs.

Wave Energy Trailblazers:

Islay LIMPET (Scotland): A pioneer in wave power technology, paving the way for future innovations.

Wave Farm (Portugal): The first multi-generator wave farm, a stepping stone towards large-scale wave energy.

The Common Thread: Turbines and Generators

Whether driven by wind, tides, or waves, these renewable energy technologies often share a common thread: turbines and generators.

Turbines: These ingenious devices transform the movement of fluids (air or water) into rotational energy.
Generators: They take the baton from turbines, converting rotational energy into the electricity that powers our world.

A Sustainable Symphony

Wind, tidal, and wave energy offer a harmonious solution to our energy needs. While challenges remain, ongoing research and development are pushing the boundaries of these technologies. As we move towards a future powered by nature, these renewable resources will play a vital role in preserving our planet for generations to come.

The Renewable Energy Revolution: A Tale of Sun, Wind, Water, and Earth

Forget dry textbooks and yawn-inducing lectures! Let's embark on an exciting journey into the world of renewable energy, where the future of our planet is being shaped.

Imagine a world where shimmering solar panels adorn rooftops like glittering jewels, where graceful wind turbines dance in harmony with the breeze, where the power of rushing rivers is harnessed for good, and where even the humble remnants of our harvests fuel our lives. This is the promise of renewable energy – a vibrant tapestry woven from the very elements that sustain us.

1. Solar Energy: A Star is Born

The sun, our celestial powerhouse, showers us with an abundance of energy every single day. Solar panels, like miniature suns themselves, capture this radiant energy and transform it into electricity. It's like having a tiny power plant right on your rooftop!

Imagine:

Sun-kissed cities: Buildings adorned with solar panels, gleaming like beacons of sustainability.
Solar-powered adventures: Electric cars zipping along highways, fueled by the sun's energy.
Empowered communities: Remote villages basking in the glow of solar-powered lights, breaking free from energy poverty.
But like any superhero, solar energy has its kryptonite:

The Night Owl: Solar power takes a break when the sun goes down, making energy storage essential.
The Land Grab: Large solar farms can sometimes compete for space with nature and agriculture.

2. Wind Energy: Catching the Breeze

Wind, the invisible force that shapes our landscapes, holds immense power. Wind turbines, with their majestic blades reaching for the sky, capture this power and convert it into clean electricity. It's like harnessing the very breath of the Earth!

Imagine:

Windswept coastlines: Offshore wind farms dotting the horizon, generating clean energy from the sea breeze.
Rural landscapes: Wind turbines standing tall amidst fields of golden wheat, powering homes and businesses.
Island paradises: Wind energy providing a lifeline to remote islands, reducing their reliance on expensive fossil fuels.

But even the mighty wind has its challenges:

The Fickle Friend: Wind can be unpredictable, making energy storage and grid integration crucial.
The Winged Wonders: Bird and bat collisions are a concern, but innovative solutions are being developed to protect these creatures.

3. Hydropower: The Power of Flow

Water, the lifeblood of our planet, has been a source of power for centuries. Hydropower harnesses the energy of flowing water, whether it's the mighty rush of a waterfall or the steady current of a river. It's like tapping into the Earth's circulatory system!

Imagine:

Mountainous marvels: Dams nestled amidst breathtaking scenery, generating clean energy while providing water storage.
Riverine rhythms: Run-of-river hydropower systems seamlessly integrated with nature, generating power without disrupting ecosystems.
Coastal communities: Tidal power harnessing the ocean's ebb and flow, providing a predictable and renewable energy source.

But even water, the essence of life, can have its drawbacks:

The Ecosystem Architect: Dams can alter river ecosystems, requiring careful planning and mitigation measures.
The Climate Conundrum: Changes in precipitation patterns due to climate change can affect hydropower generation.

4. Biomass Energy: From Waste to Watts

Biomass, the organic matter that surrounds us, holds a surprising amount of energy. From wood chips and crop residues to food waste and algae, biomass can be converted into electricity, heat, and even biofuels. It's like turning nature's leftovers into a power source!

Imagine:

Forest communities: Utilizing wood waste from sustainable forestry to power homes and businesses.
Agricultural abundance: Crop residues and animal manure transformed into bioenergy, reducing waste and providing income for farmers.
Urban renewal: Food waste diverted from landfills and converted into biogas, fueling city buses and powering homes.

But even biomass, the gift that keeps on giving, has its caveats:

The Air We Breathe: Biomass combustion needs to be carefully managed to minimize air pollution.
The Land We Share: Growing dedicated energy crops can sometimes compete with food production.

The Future is Renewable

The renewable energy revolution is not just about technology; it's about transforming our relationship with the planet. It's about harnessing the power of nature in a way that is sustainable, equitable, and beneficial for all. It's about creating a world where clean energy is abundant, affordable, and accessible to everyone.

So, join the movement, embrace the change, and let's create a brighter future powered by the sun, wind, water, and earth!

Efficiency

Efficiency: A Tale of Two Energies

Imagine energy as a mischievous sprite, always eager to transform and shift its form. Sometimes it appears as light, other times as heat, or even the movement of a machine. But this sprite is also a bit of a trickster, prone to disappearing into thin air or changing into a form we don't find useful. This is where efficiency comes in – it's our way of measuring how well we can take this sprite and guide its transformations to suit our needs.

Think of it like this: you have a bucket full of this energetic sprite (your energy input), and you want to use it to fill a smaller container (your desired output). Efficiency tells you how much of that sprite actually makes it into the smaller container and how much is spilled or escapes along the way.

Two Sides of the Same Coin

Efficiency has two main personas:

Energy Efficiency: This is like taking a snapshot of the whole process. It tells you how much of the total energy you put in ended up where you wanted it. The rest? Well, that mischievous sprite might have turned into unwanted heat or sound, or simply vanished into the surroundings.
Power Efficiency: This is more like watching a movie of the process. It tells you how much energy is being transferred per unit of time – the rate of flow, so to speak. It's like seeing how fast you can fill that smaller container from the big bucket.

The Quest for Perfection

In an ideal world, we'd capture every bit of that energetic sprite and put it to good use. But alas, no system is perfect. There's always some spillage, some transformation into unwanted forms. Friction, heat, and internal resistance are like tiny gremlins that steal away bits of our sprite.

That's why engineers and scientists are constantly on a quest to improve efficiency. They're like sprite wranglers, devising clever contraptions and methods to capture and guide as much of the sprite as possible.

Efficiency in Action

Let's see how this plays out in the real world:

The Light Bulb Revolution: Remember those old incandescent bulbs that heated up your room as much as they lit it? That's because they were terrible sprite wranglers, letting most of the energy escape as heat. Then came the LED bulbs, shining brightly while staying cool. They're much better at guiding the sprite into light, saving energy and money in the process.
The Electric Car Race: Gasoline cars are like old, clunky sprite-powered carriages, guzzling energy and spewing out a lot of wasted

heat. Electric cars, on the other hand, are sleek and efficient, converting a much larger portion of the energy into motion.
Solar Panels: Catching Sunshine: Solar panels are like sprite nets, trying to capture as many of the sun's energy sprites as possible. The more efficient they are, the more sprites they can snag and turn into electricity.

Beyond the Basics

Efficiency isn't just about everyday gadgets. It's a fundamental concept that touches every aspect of our world, from giant power plants to tiny computer chips. Scientists and engineers are constantly exploring new frontiers of efficiency, delving into advanced concepts like:

Carnot Efficiency: This is like the ultimate sprite-wrangling challenge, defining the theoretical limit of how much energy we can extract from heat.
Exergy: This is a way of measuring the quality of energy, like figuring out how much "oomph" each sprite has.
Energy Audits and Life Cycle Assessments: These are like detective investigations, tracking down energy losses and finding ways to improve efficiency throughout the entire life of a product or process.

The Efficiency Adventure

Efficiency is an ongoing adventure, a quest to harness the mischievous energy sprite and make the most of its transformations. It's a story of innovation and ingenuity, driven by the desire to optimize resources, protect our environment, and create a more sustainable future. So, the next time you flip a switch, drive a car, or use any device, remember the tale of two energies and the ongoing pursuit of efficiency.

Power

Power: The Unsung Hero of Our Dynamic World

Imagine a world where everything is frozen in time – no flickering lights, no speeding cars, no soaring rockets. This static existence would be devoid of the dynamism we experience daily, all thanks to the silent force we call power.

Power isn't some abstract concept confined to physics textbooks; it's the lifeblood of our physical world, the engine driving change. It's the reason a tiny seed can burst through the soil, a mighty river can carve canyons, and a brilliant mind can conceive groundbreaking inventions.

The Two Faces of Power

Think of power as a master juggler, effortlessly balancing two tasks:

The Workhorse: Power measures how quickly work gets done. Work, in the physics sense, involves transferring energy to an object, making it move or change. Power is all about speed – the faster the work, the greater the power.

Imagine a weightlifter hoisting a barbell. The heavier the weight and the faster they lift it, the more power they display. It's like a sprinter racing against time, where every fraction of a second matters.

The Energy Maestro: Power also gauges how swiftly energy transforms from one form to another. Energy, the capacity to do work, comes in many flavors – kinetic, potential, thermal, electrical. Power orchestrates these transformations, determining how rapidly energy shifts and flows.

Think of a blazing bonfire. The faster it consumes wood and releases heat, the higher its power. Similarly, a fast-charging phone gulps down electrical energy quicker than its slower counterpart, showcasing its superior power.

Watts, Horses, and Everyday Wonders

We measure power in watts, a tribute to James Watt, the ingenious mind behind the steam engine. One watt is like a tiny spark of power, the energy spent when a joule of work is done in one second.

But watts aren't alone; they have companions: kilowatts for household appliances, megawatts for power plants, and gigawatts for entire cities. Even horsepower, a relic from the past, still gallops alongside, reminding us of the power of horses that once drove our world.

Equations Unveiled: The Power of Relationships

The equations $P = W / t$ and $P = \Delta E / t$ aren't just mathematical symbols; they're like secret keys unlocking the mysteries of power.

They reveal that power is a balancing act:

More work in the same time means more power. A construction crane lifting a massive steel beam demonstrates immense power.
The same work in less time also means more power. A cheetah sprinting after its prey showcases explosive power.
A greater energy shift in the same time means higher power. A roaring jet engine transforming fuel into thrust exhibits tremendous power.
The same energy shift in less time also means higher power. A lightning bolt unleashing its electrical fury in a split second demonstrates nature's raw power.

Power in Action: From Pedal to Planet

Power isn't confined to equations; it's woven into the fabric of our lives:

Human Power: Cyclists pumping their legs, climbers conquering mountains, and dancers expressing emotions through movement – all embody human power.
Mechanical Power: Car engines roaring, electric motors whirring, and mighty cranes lifting heavy loads – these are testaments to mechanical power.

Electrical Power: Light bulbs illuminating our homes, refrigerators keeping our food fresh, and computers connecting us to the world – all rely on electrical power.
Renewable Power: Solar panels harnessing the sun's energy, wind turbines dancing with the wind – these are beacons of sustainable power.

Electric Vehicles: A Case Study in Power and Progress

Electric vehicles (EVs) are revolutionizing transportation, and power is at the heart of this transformation. The faster an EV accelerates, the higher the power of its electric motor. Imagine an EV zooming from 0 to 60 mph in a blink – that's power in action!

EV batteries, like energy reservoirs, store power for the journey. Fast charging technology allows these batteries to be replenished quickly, minimizing downtime and maximizing convenience.

Power: The Driving Force of Change

Power is more than just a physics concept; it's a story of change, progress, and the dynamism that shapes our world. From the smallest atom to the largest galaxy, power is the unsung hero driving the universe's grand symphony.

Understanding power empowers us to make informed choices about energy consumption, design efficient machines, and harness renewable resources, ensuring a sustainable future for generations to come. So, let's celebrate power, the engine of change that makes our world vibrant and alive.

Feeling the Squeeze: Pressure in Our World

Imagine a ballerina gracefully gliding across the stage, her weight seemingly defying gravity. Now picture a construction worker hammering a nail into a sturdy beam, each strike a burst of focused energy. What's the invisible force at play in both scenarios? It's pressure – the way force spreads itself out over an area.

Think of pressure as a tiny dancer with mighty feet. The smaller her shoes (the area), the more intensely she taps the floor (the pressure), even if her weight (the force) stays the same. This is why a sharp needle can pierce fabric with ease while a blunt object might just stretch it.

Pressure's Everyday Dance:

The Sinking Feeling: Ever walked on snow and felt your feet disappear into its fluffy embrace? That's high pressure in action! Your weight, concentrated on a small area (your shoes), overwhelms the snow's resistance. But strap on some snowshoes, and suddenly you're floating on top. Why? You've spread out the force, lowering the pressure and saving yourself from a chilly plunge.

Knife Skills 101: A dull knife is like a clumsy dancer, needing lots of force to make a cut. A sharp knife, with its tiny point of contact, is a precision instrument. It concentrates your force, creating intense pressure that slices effortlessly through whatever you're chopping.

The Tire Tale: Why do massive trucks have so many tires? It's not just for show! Those extra tires spread the truck's enormous weight, reducing the pressure on the road. Think of it as a group of dancers sharing the stage – the load is distributed, and no one gets crushed.

Pressure's Grand Performances:

Pressure isn't just about everyday objects. It's a star performer in the grand theater of the universe:

Deep Sea Drama: Dive into the ocean depths, and pressure becomes a powerful antagonist. The weight of the water above creates immense pressure, squeezing everything in its path. This is why deep-sea creatures have evolved bizarre adaptations to survive this crushing force.

Weather Wonders: The air around us is like an invisible dancer, constantly shifting and swirling. Changes in atmospheric pressure create the wind, rain, and sunshine that make up our weather. It's a never-ending performance, with pressure as the lead choreographer.

The Body's Balancing Act: Inside your own body, pressure is a vital force. Your blood pressure, the rhythmic pulse of life, keeps your blood flowing. Your lungs, like bellows, use pressure to draw in life-giving oxygen. Pressure is the silent conductor of your body's symphony.

Beyond the Basics:

Pressure isn't always a simple equation. Sometimes it's a complex dance with swirling movements and hidden forces:

Fluid Dynamics: Imagine a river rushing over rocks, or the wind whipping through a canyon. In these dynamic environments, pressure changes with the flow, creating eddies and currents. It's a fluid, ever-changing dance.

Material World: Pressure can transform materials in surprising ways. Deep within the Earth, immense pressure creates diamonds, the hardest substance known to humankind. In laboratories, scientists use high pressure to create new materials with amazing properties.

The Final Bow:

Pressure is a fundamental force that shapes our world, from the smallest atom to the largest galaxy. It's a dancer, a sculptor, a life-giver. By understanding its intricate steps, we can unlock the secrets of the universe and create a better world for ourselves.

Feeling the Pressure? Let's Dive Deep!

Imagine a ballerina gracefully balancing on the tip of her pointe shoe. That incredible feat is possible because of pressure – the force she exerts is concentrated on a tiny area, creating enough force per unit area (pressure!) to support her weight. Now, imagine her trying to balance on a fluffy pillow. Not so easy, right? That's because the same force is spread over a much larger area, resulting in lower pressure.

Pressure: It's All About Direction

Pressure isn't just about how much force is applied; it's also about the direction of that force. Think of a playful water fight with a friend. When you squeeze a water bottle, the water doesn't just dribble out; it shoots out in a jet perpendicular to the bottle's surface. That's because pressure exerts a force at a right angle to the surface it's acting on.

DIY Science: The Leaky Bottle Experiment

Want to see this in action? Grab an empty plastic bottle, a pin, and some water. Fill the bottle, poke a few holes at different spots, and watch the water jets erupt! No matter where you poke the hole, the water will always shoot out perpendicular to the bottle's surface. It's like the water is saying, "I'm going my own way!"

Nature's Barometer: The Mercury Marvel

Ever wondered how we measure the weight of the air around us? Enter the liquid barometer, a clever invention that uses mercury's density to gauge atmospheric pressure. Picture a glass tube filled with mercury, flipped upside down in a dish of mercury. The atmosphere pushes down on the mercury in the dish, supporting a column of mercury in the tube. The higher the air pressure, the taller the mercury column. It's like a balancing act between the atmosphere and the mercury!

Altitude, Temperature, and Weather: The Barometer's Best Friends

But here's the twist: atmospheric pressure isn't constant. It changes with altitude (think mountaintops vs. sea level), temperature (mercury likes to dance when it's warm!), and weather patterns (stormy skies mean lower pressure). So, a barometer is like a weather whisperer, giving us clues about the atmospheric mood swings.

From Mercury to Modern Marvels

While mercury barometers are classics, we've got fancier gadgets these days. Aneroid barometers use a sealed metal chamber that expands and contracts with pressure changes, like a tiny lung sensing the air's weight. These barometers are compact, portable, and don't involve any messy mercury spills!

Beyond the Weather Forecast

Barometers aren't just for predicting rain or shine. Pilots rely on them to determine altitude and navigate safely, while scientists use them in experiments to understand the world around us. So next time you see a barometer, remember it's not just a weather gadget; it's a window into the fascinating world of pressure!

Feeling the Squeeze: A Human's Guide to Pressure Underwater

Ever jumped into a pool and felt that weird pressure in your ears as you went deeper? That's not just you! It's the weight of the water above you pushing down. The deeper you go, the more water there is, and the stronger that push becomes. Scientists call this "hydrostatic pressure," and it's a pretty big deal, not just for swimmers, but for all sorts of things like submarines, weather patterns, and even your own blood pressure!

The Deeper You Dive, the Higher the Pressure

Think of it like this: imagine stacking pillows on top of yourself. One pillow isn't so bad, but ten pillows? You'd definitely feel the weight! It's the same with water. Each layer of water adds to the pressure pushing down on you. And just like denser pillows filled with heavier stuff would squish you more, denser liquids like seawater create more pressure than freshwater.

The Pressure Equation: A Simple Formula

Scientists have a simple formula to figure out this pressure:

Pressure = Density x Gravity x Depth

Density: How tightly packed the molecules are in a liquid. Think of honey (dense) versus water (less dense).
Gravity: That invisible force that keeps us on Earth (and makes cannonballs sink!).

Depth: How far down you go.

The Hydrostatic Paradox: Shape Doesn't Matter!

Here's a mind-bending fact: the pressure at the bottom of a container filled with liquid only depends on how high the liquid is, not the shape of the container. So, whether it's a skinny tube or a wide bowl, if the liquid level is the same, the pressure at the bottom is equal. Weird, right?

Pascal's Principle: Pressure Plays Fair

Imagine squeezing a tube of toothpaste. The pressure you apply at one end travels all the way through the toothpaste, right? That's Pascal's Principle in action! This idea is used in all sorts of cool technology, like hydraulic lifts that can hoist cars with ease.

Buoyancy: Why We Float (Sometimes)

Ever wondered how ships stay afloat? It's all thanks to buoyancy, a force that pushes upwards against submerged objects. The more water an object pushes aside (displaces), the stronger the buoyant force. This is why a giant cruise ship can float, while a tiny pebble sinks.

Real-World Examples: Pressure in Action

Deep-sea Exploration: Those incredible submersibles that explore the deepest parts of the ocean have to be super strong to withstand the crushing pressure.
Blood Pressure: Doctors use the principles of hydrostatic pressure to measure your blood pressure, making sure your heart is pumping just right.
Weather Forecasting: Changes in atmospheric pressure help meteorologists predict sunny days or stormy weather.
Hydraulic Fracturing: "Fracking" uses high pressure to crack rocks and extract natural gas, but it can also cause environmental problems.

The Bottom Line: Pressure is Everywhere!

Hydrostatic pressure is a fundamental force of nature that affects us all. From the depths of the ocean to the blood flowing through our veins, understanding pressure helps us understand the world around us and create amazing technologies. So next time you take a dip in the pool, remember the incredible science at play beneath the surface!

Thermal physics

Kinetic particle model of matter

States of matter

Imagine a world where:

Solids are like stubborn old men, refusing to change their ways or budge an inch. They're set in their routines, happy with their defined shapes and volumes. Think of a grumpy grandpa sitting in his favorite armchair, unwilling to move for anything less than a natural disaster.

Liquids are like playful children, always on the move and exploring their surroundings. They adapt to any container they're put in, taking on new shapes while maintaining their playful, fluid nature. Picture a group of kids splashing in a pool, their laughter echoing as they change their forms to make a splash.

Gases are like free-spirited artists, constantly expanding their horizons and filling every space they encounter. They're wild and untamed, refusing to be confined by boundaries or limitations. Visualize a vibrant hot air balloon soaring through the sky, expanding with the warmth of the sun and dancing with the wind.

Now, let's dive deeper into their unique personalities:

Solids: These steadfast fellows are the backbone of our world, providing structure and stability. Their tightly packed particles are like a close-knit community, holding onto each other with unwavering strength. They're the bricks that build our houses, the rocks that form our mountains, and the ice that cools our drinks.

Liquids: These adaptable adventurers are always flowing and changing, taking on new forms with every twist and turn. Their particles are like a group of friends exploring a new city, moving freely while staying connected. They're the water that quenches our thirst, the blood that flows through our veins, and the fuel that powers our vehicles.

Gases: These ethereal explorers are the nomads of the matter world, constantly moving and expanding into uncharted territory. Their particles are like independent travelers, roaming freely and exploring the vast expanse of their container. They're the air we breathe, the wind that whispers through the trees, and the steam that rises from a hot cup of coffee.

But what happens when these personalities clash?

Melting: Imagine a grumpy old man (solid) slowly warming up to a playful child (liquid) on a hot summer day. The ice cream in his hand melts, symbolizing his transformation from a rigid solid to a flowing liquid.

Freezing: Picture a group of playful children (liquid) settling down for a nap on a cold winter night. The water in the puddle freezes, representing their transition from a flowing liquid to a rigid solid.

Vaporization: Visualize a free-spirited artist (gas) escaping from a crowded room (liquid) to find their own space. The steam rising from a boiling kettle symbolizes their liberation from the confines of the liquid state.

Condensation: Imagine a group of independent travelers (gas) gathering around a campfire on a chilly evening. The dew drops forming on the grass represent their reunion as they condense from a gas to a liquid.

In conclusion, solids, liquids, and gases are like characters in a grand play, each with their own unique personalities and roles to fulfill. Their interactions and transformations shape the world around us, from the melting of glaciers to the formation of clouds. By understanding their distinct properties, we can appreciate the beauty and complexity of the matter that makes up our universe.

Particle model

Imagine a bustling city. The people within it are like the particles that make up all matter in the universe.

Solids? Think of a packed rush hour subway train. Everyone's crammed together, vibrating with impatience (that's their kinetic energy!), but basically stuck in place. They can jostle and wiggle, but they can't really go anywhere. That's a solid – particles locked in a tight embrace, giving it a definite shape and volume.

Liquids? Now imagine those same people at a crowded concert. They can move around, groove to the music, and even swap places with their neighbors, but they're still pretty close together. That's a liquid – particles with a bit more freedom to move, taking the shape of their container (the concert hall).

Gases? Finally, picture those people at a huge park on a sunny day. They're spread out, running around, playing frisbee, with tons of space between them. They can go wherever they please. That's a gas – particles with loads of energy, zipping around freely and filling up any space they find.

Temperature is like the DJ at this particle party. The hotter the music, the wilder the dancing. Turn up the heat, and those solid particles in the subway car start vibrating so hard they break free and melt into a liquid (think mosh pit!). Crank it up even more, and they'll evaporate into a gas, escaping the confines of the train car altogether.

But what happens when the DJ chills the music? As the temperature drops, the particles slow down. Eventually, if it gets cold enough, they'll barely move at all, reaching a state of minimum energy – that's absolute zero, the ultimate chill-out zone!

Think of water – the ultimate shape-shifter. As ice, it's a rigid solid, its molecules locked in a crystalline structure. Heat it up, and it melts into liquid water, flowing freely. Heat it even more, and it transforms into steam, an invisible gas that expands to fill any space.

This particle picture helps us understand why bridges expand on hot days (those particles need more room to dance!), why ice cubes melt in your drink (those solid particles can't resist the party!), and why balloons shrink in the cold (those gas particles lose their energy and huddle closer).

So, there you have it – the particle picture of matter, a universe of tiny dancers constantly moving and grooving to the beat of temperature. It's a picture full of energy, motion, and the endless transformations of matter.

Imagine a Dance Floor Packed with Tiny Dancers

Forget stuffy labs and complicated equations. Picture a crowded dance floor, pulsing with energy. Those dancers, bumping and swirling around? They're like the tiny gas particles in a container! The more dancers there are, the more crowded it gets, and the more they bump into each other and the walls. That's like pressure – the force of all those collisions spread out over the dance floor.

Cranking Up the Heat

Now, imagine the DJ puts on an upbeat track. The dancers go wild, moving faster and with more energy. They collide more often, and with more force. That's what happens when you increase the temperature of a gas. The particles get more energetic, move faster, and BAM! – higher pressure.

Shrinking the Dance Floor

Suddenly, the club owner decides to rope off half the dance floor. Now the same number of dancers are crammed into a smaller space. Collisions happen more often, and the pressure intensifies. That's Boyle's Law in action – shrink the space, and the pressure goes up!

Let's Talk Balloons and Pressure Cookers

Think of a hot air balloon. Heat the air inside, and those tiny dancers (air molecules) go wild, needing more space to move. The balloon expands, and voila! – it floats because the hot air inside is less dense than the cool air outside.

Or picture a pressure cooker. As the temperature rises, the steam molecules inside get super energized, bumping into the walls with more force. This builds up pressure, allowing food to cook faster. It's like the dancers are so wild they're practically bouncing off the walls!

Gas Laws: The Rules of the Dance Floor

Gay-Lussac's Law: Fixed dance floor, hotter music = wilder dancing and more pressure.
Charles's Law: Constant pressure, hotter music = the dancers need more space, so the dance floor expands.
Boyle's Law: Constant temperature, smaller dance floor = more crowded, more collisions, higher pressure.

From Deep Sea Diving to Brewing the Perfect Coffee

Understanding gas pressure isn't just for scientists. It's at play when scuba divers descend into the depths, ensuring they equalize pressure to avoid injury. It's behind the fizz in your soda and the perfect crema on your espresso. It's even crucial for understanding weather patterns and climate change.

So next time you encounter gas pressure, forget the textbooks and imagine those tiny dancers. It's a much more fun – and memorable – way to understand this fundamental force of nature.

Boyle's Law: The Gas That Likes to Play Squeeze

Imagine a mischievous gas, trapped inside a container. This gas loves to play a game of squeeze, and it has a special rule: the tighter you squeeze it (smaller the space), the harder it pushes back (higher the pressure). This is the essence of Boyle's Law, a fundamental principle in the world of gases.

The Equation: A Balancing Act

Boyle's Law can be summed up in a simple equation: $p_1V_1 = p_2V_2$. Think of it as a seesaw, where pressure (p) and volume (V) are always trying to balance each other out. If one goes up, the other must come down.

The Hyperbola: A Curveball of Pressure

If you were to draw a graph of this pressure-volume relationship, you'd get a graceful curve called a hyperbola. It's like a visual representation of the gas's playful struggle. As the volume shrinks, the pressure shoots up, and the curve climbs steeply. But when the volume expands, the pressure eases off, and the curve gently descends.

Real-Life Shenanigans: Boyle's Law in Action

This playful gas isn't just confined to textbooks and labs. It's all around us, pulling the strings in many everyday scenarios:

Breathing: Our lungs are like balloons that follow Boyle's Law. When we inhale, our lungs expand, the pressure inside drops, and air rushes in. When we exhale, the opposite happens.
Syringes: Pulling the plunger creates more space, lowering the pressure and sucking in the liquid. Pushing the plunger does the reverse, squeezing the liquid out.
Balloons: Ever noticed how a balloon feels firmer the more you inflate it? That's Boyle's Law at work. More air crammed into the same space means higher pressure.
Scuba Diving: Deep underwater, the immense pressure squeezes the air in a diver's lungs. Understanding Boyle's Law is crucial for safe diving.
Aerosol Cans: Pressing the nozzle releases the compressed gas within, which expands rapidly and pushes out the product.

Beyond the Basics: The Gas Gets Real

While Boyle's Law is a great starting point, real gases can be a bit more complex. They have tiny volumes of their own and sometimes even get a bit "sticky" with each other. But even with these quirks,

Boyle's Law remains a cornerstone of understanding how gases behave.

In Conclusion: A Playful Principle

Boyle's Law, with its simple equation and playful analogy, reveals a fundamental truth about the world of gases. It's a principle that pops up in countless scenarios, from the air we breathe to the depths of the ocean. So next time you encounter a gas, remember its playful nature and the delicate balance between pressure and volume.

Thermal properties and temperature

Thermal expansion of solids, liquids and gases

Imagine this:

You're holding a balloon on a frosty winter morning. It's shriveled, almost lifeless. But as you bring it inside, close to the fireplace, something magical happens. It plumps up, dances in the warm air, full of life and vibrancy. That, my friend, is thermal expansion in its simplest, most delightful form.

The Dance of the Molecules

Think of matter as a bustling dance floor. In solids, the dancers (atoms) are tightly packed, swaying gently to a slow rhythm. Heat it up, and the music gets faster, the dance more energetic. They need more space, they push outward, and the whole dance floor expands.

Liquids? A bit more freedom. Imagine a lively salsa party. Heat turns up the tempo, the dancers twirl faster, bumping into each other more often. They need a bit more room to swing those hips, and the dance floor expands again.

Now, gases – a wild, free-for-all mosh pit! Molecules bouncing off each other, off the walls, pure chaos. Add heat? It's like someone spiked the punch. The moshing intensifies, the collisions become wilder, and the walls of the mosh pit (the container) are pushed outward. Expansion!

The Two Worlds of Temperature

We measure this heat-driven dance with two different thermometers. Celsius, the everyday one, uses water as its guide. Zero is when water freezes, 100 when it boils.

But Kelvin, ah, that's the philosopher. It starts at the ultimate chill – absolute zero, where the music stops, and all motion ceases. A theoretical place so cold, it makes the Arctic feel like a sauna.

Thermal Expansion: The Good, the Bad, and the Ugly

This expansion can be a powerful force. It lifts hot air balloons, bends metal strips to flick switches in your thermostat, and even helps fit tight metal parts together.

But it's a double-edged sword. Bridges and buildings? They need breathing room, expansion joints to handle the swelling and shrinking of summer and winter. Railroad tracks? Too close together on a scorching day, and you've got a recipe for a buckled mess.

And let's not forget the delicate dance of precision. Clocks, scientific instruments – they hate this expansion. It throws off their delicate balance, making them less accurate.

In Conclusion

Thermal expansion – it's the invisible hand shaping our world, from the smallest atom to the largest skyscraper. It's a reminder that even the most solid-looking things are in constant motion, responding to the rhythm of heat. So next time you see a bridge, a balloon, or even a simple thermometer, remember the dance of the molecules, and marvel at the power of this invisible force.

Specific heat capacity

Feeling the Heat: A Dance of Molecules

Imagine a bustling dance floor, packed with energetic dancers. Some twirl gracefully, others bounce with enthusiasm, and a few zips around with boundless energy. This lively scene mirrors the invisible world of molecules within any substance. Just like our dancers, these tiny particles are in constant motion, vibrating, rotating, and colliding with each other. The more energetic the dance, the hotter the dance floor – and the higher the temperature of the substance.

Temperature: The Beat of the Molecular Dance

Temperature is like the DJ's music, setting the pace for the molecular dance. A high temperature cranks up the volume, making the molecules move faster and collide more frequently. A low temperature slows things down, with molecules swaying gently and bumping into each other less often.

But temperature isn't just about speed; it's about the average energy of the dancers. Think of a tiny sparkler compared to a bathtub full of warm water. The sparkler's molecules are moving with incredible speed, giving it a scorching temperature. The bathtub's molecules are more relaxed, resulting in a lower temperature, even though there are many more of them.

Internal Energy: The Whole Party

Internal energy is like the total energy of the entire dance party – not just the dancers' movement, but also the lights, the sound system, and the energy stored in the building itself. In a substance, internal energy includes the kinetic energy of the moving molecules, plus the potential energy stored in their bonds and interactions.

Kinetic Energy: The Dancers' Moves

Kinetic energy is all about movement. It's the energy of the molecules as they shake, spin, and zoom around. The faster they move, the more kinetic energy they have. This energy comes in different flavors:

Translational kinetic energy: Like dancers moving across the floor.
Rotational kinetic energy: Like dancers spinning in place.
Vibrational kinetic energy: Like dancers shimmying to the beat.

Heat Transfer: Sharing the Energy

Imagine someone opening a door to a cooler room. The dancers near the door will start to lose some of their energy to the cooler air, while the people in the cooler room will gain energy. This is similar to heat transfer – energy flows from a hotter object to a colder one until they reach a comfortable balance.

Phase Transitions: Changing the Dance Floor

Sometimes, the DJ changes the music, and the entire dance floor transforms. Imagine the dancers suddenly pairing up for a waltz, or forming a conga line. These changes are like phase transitions – when a substance melts, freezes, or boils. During these transitions, energy is used to rearrange the molecules, like changing partners or forming new dance formations, rather than making them move faster.

Real-World Examples

Thermometers: Like a dance floor thermometer, these devices measure the "heat" of the molecular dance, telling us the temperature.
Internal Combustion Engines: These engines harness the energy of "exploding" molecules, like a sudden burst of wild dancing, to power our cars.
Refrigerators: These appliances use refrigerants to "steal" some of the heat energy from the inside, like a cooling breeze across the dance floor.
Climate Change: The increase in Earth's average temperature is like turning up the heat on the entire planet, causing widespread changes to our global "dance floor."

Beyond the Basics

The relationship between temperature, internal energy, and kinetic energy is a complex and fascinating one. Scientists use advanced concepts like the kinetic theory of gases, statistical mechanics, and thermodynamics to understand this relationship in even greater detail. These fields help us explain everything from the behavior of individual molecules to the large-scale energy transformations that drive our world.

Imagine a dance floor. Not just any dance floor, but one made of different materials – sleek metal, rustic wood, shimmering water. Now, imagine the music starts. Heat energy, in the form of vibrant, pulsating rhythms, floods the floor. How do our dancers – the atoms within these materials – react?

That, my friend, is where specific heat capacity comes in. It's like each dancer's unique groove, their individual response to the music. Some materials, like water, are graceful, taking their time to heat up and cool down. They have a high specific heat capacity, holding onto that heat energy for dear life. Metals, on the other hand, are like fiery tango dancers, quick to react and quick to let go, showcasing a low specific heat capacity.

Think of it like this:

Water: Imagine trying to boil a pot of water for pasta. It takes a good while, right? That's because water has a high specific heat capacity, stubbornly clinging to the heat you're giving it. This makes water a fantastic coolant, whether it's in your car's radiator or keeping your body temperature stable.
Metals: Now picture a frying pan on the stove. It heats up in a flash! That's because metals have a low specific heat capacity, readily absorbing and releasing heat. This makes them perfect for cooking, but also for things like heat sinks in your computer, which whisk away excess heat.

But how do we measure this "dance fever" of the atoms?

Scientists have devised clever ways to measure specific heat capacity, like:

The "Mix and Mingle" Method: Imagine two groups of dancers at a party – one hot and energetic, the other cool and calm. When they mix, the energetic dancers lose some of their energy, while the calm ones

gain some. By measuring the final "temperature" of the party, we can figure out how much heat each group exchanged, revealing their specific heat capacity.
The "Electric Slide" Method: Here, we give our dancers a jolt of electrical energy, like plugging them into a powerful sound system. We watch how their "temperature" rises in response to this energy boost, giving us a direct measure of their specific heat capacity.

Why does all this matter?

Understanding specific heat capacity unlocks a world of possibilities. It helps us:

Design better materials: From heat-resistant spacesuits to energy-efficient buildings, knowing how materials respond to heat is crucial.
Harness energy: Specific heat capacity plays a vital role in technologies like solar water heaters and thermal energy storage systems.
Understand our planet: Water's high specific heat capacity helps regulate Earth's climate, preventing drastic temperature swings.
So next time you feel the warmth of the sun or the chill of a winter breeze, remember the dance of atoms, each with their own unique rhythm, dictated by their specific heat capacity. It's a hidden world of energy exchange, shaping our everyday experiences and driving technological innovation.

Melting, boiling and evaporation

Imagine a dance floor packed with molecules.

Melting: The music starts, slow and steady at first. The molecules sway gently, feeling the vibe. As the beat picks up, they start to shimmy and shake. Suddenly, the bass drops! The energy is infectious, and the molecules break free from their rigid formation, swirling and twirling in a joyous, liquid frenzy.

Solidification: The DJ slows the tempo, bringing in a chill, ambient track. The molecules calm down, their movements becoming more deliberate. As the music fades to a soft hum, they find their partners and settle into a graceful, synchronized waltz, forming a solid, crystalline structure.

Boiling: The DJ switches it up, spinning a high-energy, heart-pounding anthem. The molecules are ecstatic, bouncing and leaping with unrestrained enthusiasm. Some are so energized they launch themselves into the air, escaping the dance floor entirely and becoming a free-flowing gas.

Condensation: The party's winding down, and a mellow, soulful tune fills the air. The airborne molecules, tired from their aerial acrobatics, start to descend. They rejoin the crowd, their movements becoming more grounded and fluid as they transition back to a liquid state.

Think of it like a romance novel:

Melting: The ice cube, a stoic loner, is warmed by the passionate embrace of the sun. Its icy heart melts, transforming it into a flowing, expressive liquid.

Solidification: The once lively water molecules, now weary from their adventures, yearn for stability. They find solace in each other's company, forming strong bonds and settling into a peaceful, solid existence.

Boiling: The water molecules, fueled by an insatiable thirst for freedom, rebel against their liquid confines. They break free, soaring into the atmosphere like liberated spirits.

Condensation: The adventurous water vapor molecules, having explored the vast expanse of the sky, feel a pull towards home. They reunite, their passionate embrace forming dewdrops that cling to blades of grass.

Remember: Even though the temperature stays constant during these transformations, there's a hidden energy exchange, like a secret handshake between the molecules. This "latent heat" is the key to unlocking these magical phase transitions.

Boiling vs. Evaporation: A Tale of Two Transformations

Imagine a bustling city. Boiling is like a massive protest where everyone rushes out onto the streets at once, energized and shouting. Evaporation, on the other hand, is like individuals slipping away unnoticed, one by one, seeking a quieter life elsewhere.

1. The Great Escape: Boiling vs. Evaporation

Feature	Boiling	Evaporation
Definition	A raucous party where liquid molecules, fueled by heat, break free from their bonds and rise up in bubbles, escaping into the air.	A stealthy escape where individual molecules at the surface, feeling adventurous, gather enough energy to slip away into the atmosphere.
Temperature	Happens only when the party reaches fever pitch - the boiling point.	Can happen at any temperature, even on a chilly day.
Location	The whole liquid joins the party - it's a full-blown rebellion!	Only the surface molecules get to escape - it's an exclusive club.
Speed	Fast and furious, like a flash mob.	Slow and steady, like a solo traveler.
Energy Source	Needs a fire to get the party started (external heat source).	Can use the ambient energy of the surroundings, like soaking up the sun.
Bubble Formation	Bubbles galore! It's a bubbly celebration.	No bubbles here, just a quiet departure.
Cooling Effect	The party keeps going strong, no cooling down.	As the adventurers leave, things get a bit cooler for those left behind.

Export to Sheets

2. Evaporation: The Art of Escaping

Think of a pool table. The molecules in a liquid are like the billiard balls, constantly moving and bumping into each other. Some are slowpokes, while others are speed demons.

Kinetic Energy: This is like the speed of the billiard balls. The faster they move, the more kinetic energy they have.

Intermolecular Forces: These are like the bumpers on the pool table, keeping the balls from flying off. To escape, a molecule needs enough energy to overcome these forces.
Escape Artists: The fastest molecules, those with the most energy, manage to break free from the surface and escape into the air.
Cooling Down: As the fastest molecules leave, the average energy of the remaining ones decreases, just like the pool table becomes calmer after the most energetic balls are gone.

Factors Affecting Evaporation:

Temperature: Turning up the heat is like giving the billiard balls a jolt of energy, making them move faster and escape more easily.
Surface Area: A larger surface area is like a bigger pool table, giving more molecules a chance to escape.
Intermolecular Forces: Stronger forces are like stickier bumpers, making it harder for the molecules to break free.
Humidity: High humidity is like a crowded room, making it harder for the escapees to find their way out.
Air Movement: A breeze is like a fan blowing across the pool table, carrying away the escaped molecules and making room for more.

Examples of Evaporation in Action:

Drying clothes: The water molecules escape into the air, leaving your clothes dry and ready to wear.
Cooling a cup of tea: The hottest water molecules escape, taking their heat with them and leaving your tea refreshingly cool.
Formation of clouds: Water evaporates from the Earth's surface, rises up, and forms those fluffy white clouds in the sky.
Evaporation of sweat: Your body cools down as sweat evaporates from your skin, taking heat with it.

Case Studies:

Evaporative Cooling: This is like using a fan to cool down on a hot day. Water is evaporated into a stream of air, cooling the air as it absorbs the heat.
Salt Production: Imagine leaving a shallow pool of seawater to dry in the sun. The water evaporates, leaving behind the salt.

Desalination: This is like filtering the salt out of seawater. Evaporation is used to separate the water from the salt, leaving you with fresh, drinkable water.

In Conclusion:

Evaporation is a fascinating process that plays a vital role in our world, from the clothes we wear to the water we drink. By understanding how it works, we can appreciate the intricate dance of molecules and the amazing ways that nature uses this simple process to shape our world.

Evaporation: The Great Escape

Imagine a bustling city, filled with molecules zipping and zagging, bumping into each other like hurried commuters. That's kind of what it's like inside a liquid. Now, picture the most energetic of these molecules, fueled by heat, bursting free from the city limits and soaring into the open sky. That, my friend, is evaporation!

1. The Escape Artists

Not all molecules have the guts to make a break for it. Several factors determine how many daredevils successfully escape the liquid metropolis:

Temperature: The Heat is on! Think of temperature as the energy drink of molecules. The hotter it gets, the more amped up they become, and the more likely they are to break free from their liquid bonds. A sun-drenched puddle disappears faster than one in the shade, right?

Surface Area: Room to Breathe! Imagine a crowded concert. It's much easier to slip out if you're near the edge, right? The same goes for liquids. A shallow pan of water offers more escape routes than a deep bucket, so it evaporates quicker.

Air Movement: A Breeze to Freedom! Imagine a gentle breeze whisking away those escaped molecules, making room for more. That's what air movement does. It prevents a traffic jam of water vapor above the liquid, encouraging more molecules to join the airborne party. Think of how quickly your clothes dry on a windy day!

2. The Chilling Effect

Ever notice how sweating cools you down? That's evaporation in action! As those energetic molecules escape, they take their heat with them, leaving the remaining liquid cooler. It's like the coolest kids leaving a party, bringing the overall energy level down.

3. Latent Heat: The Secret Password

To escape the liquid city, molecules need a secret password: Latent heat. This is the energy required to break the bonds holding them in liquid form. Think of it as the entry fee to the "gas phase" club. During this transition, the temperature stays the same, like a pause before the grand escape. The energy is used to break the bonds, not to raise the temperature.

Evaporation in Action

From the steam rising from a hot cup of coffee to the rain that falls from the sky, evaporation is everywhere! It shapes our weather, dries our clothes, and even keeps us cool. So next time you see a puddle disappear, remember the tiny escape artists making a break for freedom, powered by heat, surface area, air movement, and that secret password: latent heat.

Transfer of thermal energy

Conduction

Unveiling the Secrets of Heat: A Journey Through Thermal Conductivity

Imagine a world where warmth couldn't spread its comforting embrace, or where a refreshing chill couldn't penetrate the sweltering heat. This is a world without thermal conductivity, the unsung hero of our everyday experiences. Let's embark on a journey to uncover the magic behind this phenomenon, where we'll witness the dance of atoms and the flow of invisible energy.

Experiment 1: The Rod Race

Picture a lineup of contestants, each a rod made of a different material: sleek copper, sturdy aluminum, earthy wood, and crystalline glass. At the starting line, a burst of heat ignites their ends, setting off a thrilling race. As the heat surges through their bodies, we watch in anticipation. Who will emerge as the champion of heat transfer?

Experiment 2: The Waxing and Waning of Heat

In this challenge, our contestants are adorned with delicate wax hats, each topped with a tiny flag. As the heat intensifies, the wax begins to melt, threatening to topple the flags. Which rod will lose its flag first, succumbing to the heat's relentless advance?

Experiment 3: The Ice Cube's Demise

Atop each contestant's head rests a glistening ice cube, a symbol of coolness and resilience. But as the heat encroaches, the ice cubes begin their slow surrender, transforming into shimmering droplets. Which contestant will witness the quickest demise of its icy companion?

Experiment 4: The Heat Meter's Verdict

In this final test, a sophisticated heat meter takes center stage, meticulously measuring the flow of heat through each contestant's body. The results are in, revealing the true champions of thermal conductivity.

Delving Deeper: The Microscopic World of Heat Transfer

Let's peer into the hidden realm of atoms and molecules, where the true magic of thermal conductivity unfolds. In solids, heat travels through the vibrations of these tiny particles, like a whisper passing through a crowd. In metals, an army of free electrons joins the fray, carrying heat with unparalleled speed.

Examples from Nature and Technology

From the fiery depths of the Earth to the sleek surfaces of our smartphones, thermal conductivity plays a vital role in our world. Diamonds, with their flawless crystal structure, are masters of heat transfer, while aerogels, with their airy composition, are champions of insulation.

The Future of Heat Control

Scientists are constantly pushing the boundaries of thermal conductivity, exploring new materials and techniques to harness this powerful force. Imagine a world with self-cooling devices, super-efficient insulation, and even thermal cloaking devices that can render objects invisible to heat.

Conclusion

Thermal conductivity is a symphony of atomic vibrations and electron movements, a hidden force that shapes our world in countless ways. By understanding its secrets, we can unlock new possibilities for energy efficiency, thermal management, and even invisibility to heat.

Convection

Imagine a dance floor packed with people. The music starts, and those closest to the speakers feel the energy first. They begin to move, bumping into others and creating a ripple effect. This is similar to how convection works in liquids and gases.

Think of heat as an energetic dance instructor. It excites the molecules, making them move faster and spread out. This creates a less dense area, like a pocket of space opening up on the dance floor.

Now, imagine a disco ball casting its light. The less dense, "hot" molecules are like beams of light, rising upwards. As they rise, they cool down—the music fades, and the dancers slow down. This makes them denser, like partygoers moving towards the quieter edges of the dance floor. Eventually, they sink back down, ready to be energized again.

This constant cycle of rising and sinking is a convection current. It's like a never-ending dance, driven by the energy of heat and the interplay of density.

Let's see this in action with some fun experiments:

The Purple Potion: Imagine dropping a magical purple crystal into a cauldron of water. As you heat the cauldron, the purple potion starts to swirl and rise in mesmerizing patterns, revealing the hidden dance of convection. (This is the potassium permanganate experiment.)

The Smoky Stage: Picture a miniature stage with two tiny candles as spotlights. When you introduce a wisp of smoke, it dances and twirls, carried by the rising heat from the candles. It's like a theatrical performance showcasing the movement of convection currents in the air. (This is the smoke box experiment.)

The Floating Lantern: Imagine a delicate lantern, filled with the warmth of a thousand fireflies, gently ascending into the night sky. This is like a hot air balloon, where the heated air inside makes it lighter than the surrounding air, allowing it to float effortlessly. (This is the hot air balloon experiment.)

Convection is all around us:

The Breath of the Earth: Convection drives weather patterns, creating winds that whisper across the land and ocean currents that carry marine life on grand adventures.
The Earth's Inner Dance: Deep within our planet, convection currents in the mantle are like a slow, powerful waltz, driving the movement of continents and shaping the Earth's surface.
The Comfort of Home: On a chilly evening, a convection heater acts like a cozy fireplace, warming the air and creating a comforting flow of warmth throughout your home.
The Culinary Symphony: In a convection oven, hot air circulates like a conductor's baton, ensuring that your food is cooked to perfection with every note of flavor.
By understanding the dance of convection, we can appreciate the intricate ways in which heat and density interact to create the world around us. It's a captivating performance that plays out on scales both grand and intimate, from the depths of the Earth to the warmth of your morning coffee.

Radiation

Feeling the Heat: A Tale of Infrared Radiation

Imagine a bustling city on a hot summer day. The sun beats down, its invisible rays warming the streets, buildings, and people below. This warmth you feel, this invisible energy traveling through space, is infrared radiation. It's the same kind of heat you feel when you cozy up next to a fireplace or stand close to a hot stove.

Now, picture tiny dancers, atoms and molecules, constantly moving and shaking within every object around you. The hotter the object, the wilder the dance. These energetic dancers release their excitement as infrared radiation, like invisible sparks flying off into the air.

This radiant energy, like a wave in the ocean, travels through the vastness of space, carrying warmth from the sun to our planet. Unlike sound waves that need air to travel, infrared radiation can journey through the emptiness of the cosmos, bringing us the sun's warmth from millions of miles away.

When these invisible waves encounter an object, they can be absorbed, much like a sponge soaking up water. The object's atoms and molecules catch these waves, their dance becoming more vigorous, and the object itself grows warmer.

But not all objects are equal in this dance of heat. Dark-colored objects, like a black car on a sunny day, are eager to join the party, absorbing more infrared radiation and heating up faster. Rough surfaces, like a brick wall, offer more space for the radiant energy to grab onto, making them better absorbers than smooth, shiny surfaces.

Even the air around us plays a role in this invisible dance. Greenhouse gases, like carbon dioxide and water vapor, act like a blanket, trapping infrared radiation and keeping our planet warm. This natural process, the greenhouse effect, is essential for life on Earth, but too much of a good thing can lead to our planet overheating.

Infrared radiation isn't just about feeling warm; it has many fascinating applications. Infrared cameras, like special eyes that can see heat, are used in medicine to detect inflammation, in industry to find hidden flaws in machinery, and even in search and rescue missions to locate people in the dark.

So, the next time you feel the warmth of the sun on your skin or the heat radiating from a fire, remember the invisible dance of infrared radiation, a silent symphony of energy constantly shaping our world.

Imagine a Tiny Dancer...

...a tiny, jiggling dancer, twirling and shaking with incredible energy. This dancer is an atom, and its frantic dance is what we perceive as heat. The hotter the object, the wilder the dance, and the more this tiny dancer bumps into its neighbors. These collisions generate ripples of energy that radiate outward – that's thermal radiation!

Think of it like this: every warm object is a stage filled with these tiny dancers, all emitting their unique dance vibrations. These vibrations travel as electromagnetic waves, invisible to our eyes but felt as heat. The hotter the object, the more energetic the dance, and the more intense the radiation.

Two Factors Influence the Intensity of the "Dance Party":

Temperature: Just like a feverish dancer moves with more intensity, hotter objects radiate more energy. In fact, doubling the temperature doesn't just double the radiation, it increases it by a factor of sixteen! It's like turning up the music and watching the dance floor explode with energy.

Surface Area: Imagine our dancers now performing on a bigger stage. More dancers mean more vibrations, and a larger surface area allows more of those vibrations to escape as radiation. Think of a cozy fireplace versus a tiny candle flame – the fireplace radiates more heat because it has a larger surface area.

Let's Play with Some Experiments!

Experiment 1: The Colorful Cans

Grab two identical cans, one painted black and the other white. Fill them with hot water and watch as the black can cools down faster. Why? Because black surfaces are like expert dance instructors, absorbing the "dance moves" (radiation) and emitting them with greater efficiency.

Experiment 2: The Melting Wax Mystery

Coat the bottom of two cans, one black and one white, with wax. Place them under a heat lamp and witness the wax on the black can melt first. The black can's superior absorption skills turn it into a radiant dance floor, quickly transferring the heat to the wax and melting it away.

Thermal Radiation in Action:

Sunbathing Lizards: Dark-colored lizards bask in the sun, their skin efficiently absorbing the warmth to keep them energized.

Cozy Blankets: Snuggling under a blanket? It's trapping the heat radiating from your body, creating a warm cocoon.
Thermos Magic: A thermos keeps your coffee hot by using a reflective inner surface, preventing those tiny dancers from escaping with your precious heat.
The world is a stage for these tiny dancers, and understanding their moves helps us understand how heat travels and how we can harness its power. From designing efficient solar panels to keeping our homes warm, thermal radiation plays a vital role in our everyday lives.

Consequences of thermal energy transfer

Everyday Heat Transfer: A Creative Journey

Imagine heat as an energetic traveler, always seeking new destinations and adventures. This traveler has three distinct modes of transport: conduction, convection, and radiation. Let's embark on a creative journey to explore how these modes of transport shape our everyday experiences.

Conduction: The Molecular Handshake

Picture a bustling city where molecules are the citizens, constantly interacting and exchanging information. When heat arrives in this city, it's like a wave of excitement that spreads through handshakes. The warmer molecules, energized by the heat, shake hands with their cooler neighbors, transferring their energy and raising the overall temperature.

In the Kitchen:

The Cast Iron Maestro: This seasoned conductor orchestrates a symphony of heat distribution, ensuring your steak is seared to perfection and your stew simmers evenly.
The Copper Virtuoso: A master of rapid temperature changes, the copper pan allows you to sauté vegetables with finesse and craft delicate sauces.
The Nonstick Illusionist: With its slippery surface, this pan prevents culinary mishaps, but beware of overheating its delicate coating.

Convection: The Thermal Updraft

Imagine a hot air balloon festival, where colorful balloons rise gracefully into the sky. This is convection in action, where warm air, less dense than its surroundings, ascends like a balloon, carrying heat along with it.

Heating Your Home:

The Radiant Radiator: Like a warm hug on a chilly day, the radiator gently warms the surrounding air, creating a cozy atmosphere.
The Fireplace Ballet: Flames dance and flicker, drawing in cool air like an audience captivated by the performance, while warm air rises like a graceful ballerina.
The Forced Air Express: A powerful fan propels heated air through a network of ducts, delivering warmth to every corner of your home.
Radiation: The Energetic Messenger

Imagine the sun, a celestial beacon radiating warmth and light across vast distances. This is radiation, where heat travels as electromagnetic waves, needing no physical medium to carry its message.

Everyday Encounters:

Sunlight's Embrace: Bask in the sun's radiant glow, feeling its warmth penetrate your skin, a reminder of its life-giving power.
Microwave Magic: Invisible waves penetrate your food, exciting water molecules and transforming a cold meal into a steaming delight.
Infrared Comfort: Feel the soothing warmth of an infrared heater, its rays targeting your body directly, providing instant comfort.

Conclusion:

Heat transfer is an invisible force that shapes our world in countless ways. By understanding its three modes – conduction, convection, and radiation – we can appreciate the intricate dance of energy that occurs all around us. From the kitchen to our homes, heat transfer is an essential part of our everyday lives, providing comfort, sustenance, and a deeper understanding of the universe.

Feeling the Heat: Everyday Adventures in Conduction, Convection, and Radiation

Imagine this: You wake up on a chilly winter morning. You reach for your trusty infrared thermometer – a sleek, futuristic device that measures your temperature with a beam of invisible light. No more fumbling with those old-school thermometers! This marvel of modern technology relies on radiation, the same process that warms you by the fireplace. It's like a tiny heat detective, sensing the infrared waves emitted by your forehead and revealing your temperature in a flash.

Now, you're craving a hot cup of coffee. You pour steaming hot coffee into your trusty travel mug, a marvel of thermal engineering. This ingenious invention keeps your coffee piping hot for hours, thanks to its clever use of insulation. Inside that sleek exterior, a vacuum-sealed layer prevents heat from escaping through conduction (direct contact) or convection (airflow). It's like a mini-winter wonderland for your coffee, keeping it cozy and warm while the outside world remains chilly.

As you step out the door, you bundle up in your warmest winter coat. The fluffy down filling traps a layer of warm air close to your body, preventing heat loss through conduction and convection. It's like wearing a personal, portable sauna, keeping you snug and toasty even in the fiercest blizzard.

Later in the day, you decide to bake some cookies. As you slide the tray into the oven, you witness convection in action. The hot air inside the oven circulates, distributing heat evenly and baking your cookies to golden perfection. Meanwhile, the radiant heat from the oven's heating elements toasts the cookie tops, giving them that irresistible crispy texture.

These are just a few examples of how heat transfer shapes our everyday lives. From the infrared thermometers that keep us healthy to the cozy coats that keep us warm, conduction, convection, and radiation are constantly at work, making our lives more comfortable and convenient. So next time you feel the warmth of the sun on your face or enjoy a hot cup of cocoa on a cold winter's day, take a moment to appreciate the fascinating science of heat transfer!

Waves

General properties of waves

Waves: Nature's Dance of Energy

Imagine the universe as a vast, interconnected dance floor, where energy is the music and waves are the dancers. These ethereal dancers don't move the floor itself, but they transmit the rhythm and vibrancy of the music across the expanse. From the gentle sway of ocean tides to the vigorous shaking of an earthquake, waves are the universe's way of expressing its energy.

1. The Ghostly Transfer of Energy

Think of a whisper traveling across a crowded room. The words move from mouth to ear without any of the people in between shifting from their spots. This is the magic of waves – they transport energy without displacing matter.

Picture a pebble creating ripples in a pond. The water molecules don't journey with the ripples; they merely dance in circles, returning to their original positions once the wave passes. Similarly, sound waves make air molecules vibrate, passing on the sound without creating a breeze. Even light, the cosmic messenger, travels as a wave, carrying its radiant energy without any physical substance hitching a ride.

2. The Choreography of Waves

Waves are like choreographed dances, each with its unique steps and patterns. Let's peek into their rehearsal room:

2.1 Ropes and Springs: The Line Dancers

Ropes: Imagine a jump rope held taut between two friends. A flick of the wrist sends a wave rippling down the rope, making its fibers dance perpendicular to the wave's direction. This is a transverse wave, like a snake slithering across the floor.

Springs: Springs are like coiled dancers, their movements creating longitudinal waves. When you squeeze a spring, you create compressed regions (compressions) and stretched regions (rarefactions), making the coils dance back and forth along the spring's length.

2.2 Water Waves: The Ripple Tank Ballet

A ripple tank is like a miniature stage for water waves. A light shine from above, casting shadows of the waves on a screen below, revealing their graceful ballet:

Wavefronts: Imagine lines connecting the dancers in a chorus line who are performing the same move. These lines are like wavefronts, marking the points of a wave that are in sync.

Wavelength: The distance between two consecutive "high kicks" or "low dips" in the water wave ballet is the wavelength.

Frequency: How many waves pass a fixed point in one second? That's the frequency, the tempo of the wave dance.

Amplitude: The height of a dancer's leap or the depth of their plié is like the amplitude of a wave – the maximum displacement from their rest position.

Wave speed: How fast the wave travels across the stage is its wave speed, a product of its wavelength and frequency.

2.3 Types of Waves: The Dance Styles

Transverse Waves: These are like dancers waving ribbons, their movements perpendicular to the direction the wave travels. Light waves and water waves are examples of this style.

Longitudinal Waves: Like a conga line, the dancers move back and forth along the direction of the wave. Sound waves are the masters of this dance form.

2.4 Wave Phenomena: The Special Effects

Waves have a few tricks up their sleeves:

Reflection: Like a dancer bouncing off a mirrored wall, a wave reflects when it hits a barrier, changing its direction.

Refraction: Imagine a dancer moving from a slippery floor to a sticky one, changing their speed and direction. This is refraction, like light bending as it passes from air to water.

Diffraction: When a wave encounters an obstacle, it spreads out like a dancer improvising around a prop.

Interference: When two waves meet, they can either amplify each other (constructive interference) or cancel each other out (destructive interference), like dancers coordinating their moves.

2.5 Waves in the Real World: The Grand Performances

Earthquakes: Seismic waves are the Earth's way of releasing pent-up energy, creating both transverse and longitudinal waves that ripple through its interior.

Music: Musical instruments are like wave orchestras, producing sound waves of different frequencies and amplitudes to create melodies and harmonies.

Medical Imaging: Ultrasound uses high-frequency sound waves to create images of our insides, like an echolocation dance revealing hidden structures.

Wireless Communication: Radio waves, a type of electromagnetic wave, allow us to communicate wirelessly, like sending messages through the cosmic dance floor.

2.6 The Encore

Waves are the universe's tireless performers, constantly moving and grooving, transferring energy and creating the dynamic world we experience. From the smallest ripple to the grandest cosmic wave, they remind us that the universe is a stage for an endless dance of energy.

Waves: A Symphony of Motion and Energy

Imagine yourself at the beach, the sun warming your skin as you listen to the rhythmic crashing of waves against the shore. These mesmerizing, powerful forces of nature are more than just a beautiful sight; they are a fundamental phenomenon that permeates our world, from the ocean's depths to the light that illuminates our lives.

Unveiling the Secrets of Waves

At their core, waves are disturbances that travel through a medium or space, carrying energy without transporting matter. Think of it like a whisper traveling across a crowded room – the message is passed, but the people remain in place.

Key Features: Decoding the Language of Waves

Wavefront: Picture the leading edge of a wave, where all the water molecules dance in perfect unison. This is the wavefront, a surface that connects all points in the same phase of oscillation. Wavefronts can take on various shapes, like the circular ripples spreading from a pebble tossed into a pond, or the flat plane of light emitted from a distant star.

Wavelength (λ): Imagine two surfers riding the crest of consecutive waves. The distance between them is the wavelength, the spatial separation between two corresponding points on a wave.

Frequency (f): If you count how many wave crests pass a fixed point in one second, you're measuring the frequency, a measure of how often a wave completes a cycle. Think of it as the tempo of the wave's dance.

Crest and Trough: The crest is the wave's peak, its highest point, while the trough is its valley, the lowest point. Together, they create the wave's rhythmic rise and fall.

Amplitude (A): Picture a child on a swing, soaring higher and higher. The amplitude is the maximum distance the child moves from the swing's resting point, reflecting the wave's intensity.

Wave Speed (v): Imagine a dolphin leaping through the waves. The speed at which the wave travels is its wave speed, influenced by the properties of the medium, such as the ocean's depth and currents.

Examples: Waves in Action

Sound Waves: When you listen to your favorite song, you're experiencing sound waves, longitudinal vibrations traveling through the air. The higher the frequency, the higher the pitch, and the larger the amplitude, the louder the sound.

Light Waves: These transverse waves, capable of traveling through the vacuum of space, bring us the vibrant colors of the world. Each color has its own unique frequency and wavelength, from the fiery red of a sunset to the calming blue of the sky.

Earthquake Waves: When the earth trembles, it sends out seismic waves, both longitudinal and transverse, that ripple through the planet. By studying these waves, scientists can pinpoint the earthquake's epicenter and magnitude.

Radio Waves: These invisible messengers carry our favorite radio programs and television shows, each broadcasting at a specific frequency.

Ocean Waves: The majestic waves of the ocean are a complex interplay of longitudinal and transverse waves, shaped by wind, currents, and the ocean floor.

Beyond the Basics: Delving Deeper into the World of Waves

The study of waves extends far beyond these fundamental concepts, encompassing fascinating phenomena like:

Wave Interference: When waves meet, they can either amplify each other (constructive interference) or cancel each other out (destructive interference), creating intricate patterns of energy.

Wave Diffraction: Ever wondered how you can hear music from around a corner? This is due to wave diffraction, the bending of waves around obstacles.

Wave Reflection: When a wave encounters a boundary, it can bounce back, like light reflecting off a mirror, creating an image.

Wave Refraction: As waves pass from one medium to another, they can change direction, like the apparent bending of a straw in a glass of water.

Doppler Effect: The change in frequency of a wave as the source and observer move relative to each other, explaining why an approaching ambulance siren sounds higher pitched.

Conclusion: Riding the Waves of Knowledge

By understanding the fundamental properties of waves and their intricate interactions, we unlock a deeper understanding of the world around us, from the smallest particles to the vast expanse of the universe. Waves are a symphony of motion and energy, a testament to the beauty and complexity of the natural world.

Transverse Waves: When Things Go Up and Down, But Travel Sideways

Imagine a busy city street. The cars are all zipping along in one direction, but inside, the people are bobbing their heads to music, moving up and down. That's kind of like a transverse wave! The energy (the cars) moves forward, but the actual stuff doing the waving (the people) moves perpendicular to that.

Let's Break it Down

What they are: Transverse waves are like those head-bopping passengers. They move up and down (or side to side), while the wave itself travels forward. Think of a snake slithering – its body moves side to side, but it travels forward.

The magic formula: Wave speed, frequency, and wavelength are all linked. It's like a recipe – change one ingredient, and you have to adjust the others to keep the balance. Increase the frequency (more head bobs per second), and the wavelength (distance between bobs) gets shorter, as long as the wave speed stays the same.

Where to find them: Transverse waves are everywhere! Light, those ripples in your coffee when you walk, and even the vibrations in a guitar string when you strum it.

Examples that Blow Your Mind

Electromagnetic waves: These are the rockstars of transverse waves. They're light, radio waves, X-rays – basically, the invisible stuff that makes our world go round. They're made of electric and magnetic fields that wiggle perpendicular to each other, like two dancers doing the tango while moving across the dance floor.

Water waves: Think of a surfer riding a wave. The wave moves towards the shore, but the surfer bobs up and down. That's the transverse motion in action!

Earthquake S-waves: These are the bad boys of earthquakes. They shake the ground perpendicular to their travel direction, like a dog shaking off water after a bath. And guess what? They can't travel through liquids! That's how we know the Earth's outer core is liquid – S-waves get stopped dead in their tracks.

Cool Properties

Amplitude: How big the wave is. Think of it like the height of a roller coaster – the bigger the amplitude, the wilder the ride.

Period: The time it takes for one complete wave cycle. Like, how long it takes for that head-bopping passenger to go up and down once.

Phase: Where you are in the wave cycle. Are you at the peak of the head bob, or down in the slump?

Polarization: This is a bit trickier, but think of it like this: if the wave is a jump rope, polarization is whether you're moving it up and down or side to side.

Real-World Magic

Optical fibers: They use light waves to send information, like super-fast messengers. And because light can be polarized, we can cram even more information into those tiny fibers.
Ultrasound: Those cool pictures of babies in the womb? That's sound waves, but at a frequency we can't hear. They bounce off the baby, and a computer turns those reflections into an image.
Music: When you pluck a guitar string, it vibrates in a transverse wave. The frequency of that vibration determines the note you hear. So, music is basically just playing with transverse waves!

The Adventure Continues

Transverse waves are like a hidden code to the universe. The more we learn about them, the more we understand how things work, from the tiniest atom to the biggest galaxy. So, keep exploring, keep questioning, and keep riding those waves!

Imagine a Pulse Traveling Through a Crowd

Picture yourself at a crowded concert, the music pumping, the energy palpable. Suddenly, a wave of excitement surges through the audience. People aren't moving from their spots, but they're swaying back and forth, their bodies bumping rhythmically against each other. This, my friend, is a real-life example of a longitudinal wave in action!

Longitudinal Waves: The Essence of Vibration

Longitudinal waves are all about vibrations that move in the same direction as the energy they carry. Think of sound waves traveling through the air. When a speaker vibrates, it creates zones where air molecules bunch up (compressions) and zones where they spread out (rarefactions). These zones of compression and rarefaction travel outward, carrying the sound to your ears.

Examples in Our World

The Symphony of Sound: From the sweet melodies of a violin to the roar of a jet engine, sound waves are longitudinal waves in action, creating vibrations in the air that our ears interpret as sound.

The Earth's Tremors: When earthquakes shake the ground, they send out seismic P-waves, a type of longitudinal wave that travels through the Earth's layers, causing the ground to compress and expand like a giant, trembling jelly.

Ultrasound's Gentle Touch: Medical professionals use ultrasound imaging to peer inside the human body. High-frequency sound waves, transmitted and reflected within our tissues, create images of our organs, helping doctors diagnose and treat various conditions.

Wave Behavior: A Dance of Interactions

Waves are dynamic, constantly interacting with their surroundings. They can bounce off surfaces (reflection), bend as they pass from one medium to another (refraction), or spread out as they encounter obstacles (diffraction).

Reflection: Imagine throwing a ball against a wall. It bounces back, right? That's reflection. Light bouncing off a mirror, sound echoing in a canyon - these are examples of wave reflection.

Refraction: Ever noticed how a straw seems to bend when placed in a glass of water? That's refraction - light waves changing direction as they pass from water to air.

Diffraction: Think of ocean waves bending around a pier or sound waves traveling around corners. This bending and spreading of waves are diffraction.

The Grand Finale: Waves in Action

From the music we enjoy to the medical technologies that save lives, longitudinal waves play a vital role in our world. They are the invisible messengers of energy, carrying information and shaping our experiences in countless ways.

Key Takeaways:

Longitudinal waves vibrate in the same direction as the energy they carry.
Sound waves and seismic P-waves are common examples.
Waves exhibit behaviors like reflection, refraction, and diffraction.
Understanding wave behavior allows us to harness their power for various applications.

Imagine a world where sound bends around corners like a whisper in the wind, and light dances through tiny openings, painting vibrant patterns on a screen. This is the magic of diffraction, a phenomenon that reveals the wave nature of light and sound, blurring the lines between the physical and the ethereal.

Think of waves as mischievous sprites, always seeking the path of least resistance. When they encounter an obstacle, they don't simply give up; they find clever ways to navigate around it. This is diffraction in action - the bending and spreading of waves as they encounter an obstacle or gap.

The secret to this wave wizardry lies in the relationship between the wavelength and the size of the gap. It's a delicate dance, a game of proportions where the outcome depends on who's leading the tango.

When the gap is much larger than the wavelength, the waves waltz through with minimal fuss. Imagine a grand ballroom with doors wide open, inviting the dancers to glide through effortlessly. This is the case with light passing through a doorway; it barely notices the opening and casts sharp shadows, maintaining its composure.

But when the gap narrows, approaching the size of the wavelength, the dance floor becomes more intimate. The waves, now forced to squeeze through, begin to sway and swirl, creating a mesmerizing pattern of bright and dark fringes. This is like sound waves weaving through an open door, filling the room with their melodious presence, even reaching those hidden from direct view.

And when the gap becomes too small, like a keyhole barring entry, the waves are left with no choice but to retreat. They can't squeeze through, their energy effectively blocked. This is akin to radio waves encountering a metal mesh, their passage denied, creating a sanctuary within – a Faraday cage.

To truly appreciate this dance of waves, picture a single beam of light encountering a narrow slit. As the slit narrows, the light waves, feeling constrained, spread out like a fan, painting a vibrant pattern on a screen. This is diffraction at its most playful, revealing its secrets through a mesmerizing display of light and shadow.

But diffraction isn't confined to gaps; it's equally fascinated by edges. Imagine waves gracefully curving around a sharp corner, like a river meandering through a valley. This is edge diffraction, a phenomenon that allows sound to whisper around corners and radio waves to navigate around buildings, keeping us connected even when obstacles stand in the way.

In this world of diffraction, wavelength is key. Longer wavelengths, like those of radio waves, are more adventurous, bending and spreading with ease, while shorter wavelengths, like those of visible light, are more reserved, maintaining their direction.

So, the next time you hear a sound from around a corner, or see light painting patterns on a screen, remember the magic of diffraction. It's a reminder that the world is full of hidden wonders, waiting to be discovered by those who dare to look beyond the surface. It's a dance of waves, a game of proportions, and a testament to the ingenuity of nature.

Making Waves: A Ripple Tank Adventure

Imagine a world in miniature where you can control the very fabric of the sea. No, it's not some fantastical video game, but a real, tangible tool that unlocks the secrets of waves: the ripple tank!

Think of it as a stage for water waves to perform their mesmerizing dance. A shallow, transparent tank filled with a thin layer of water becomes our theater, where light from above casts dramatic shadows of the waves onto a screen below. It's like a live action movie, but starring physics instead of Hollywood actors.

The Ripple Tank Orchestra

This watery stage wouldn't be complete without its talented performers and crew:

The Tank: Our stage, a sleek and shallow container made of glass or acrylic, ready to hold the spotlight.
Water: The star of the show, shimmering and ready to ripple to the rhythm of the waves.
Wave Generator: The conductor of our orchestra, creating the waves with a gentle dip or a more sophisticated vibration.
Light Source: The spotlight operator, illuminating the water from above and casting those captivating shadows.
Screen: The movie screen, capturing the wave action in all its glory for our viewing pleasure.

Setting the Stage for Wave Wonders

Creating this watery performance is surprisingly simple. Just follow these steps:

Steady the Stage: Place the tank on a stable surface, ensuring our performers have a solid foundation.
Fill the Sea: Gently pour water into the tank, creating a shallow ocean about 1 cm deep.
Cue the Conductor: Introduce the wave generator to the water, ready to create our wave stars.
Shine the Spotlight: Position the light source above the water, casting dramatic shadows onto the screen below.
Raise the Curtain: With the screen in place, we're ready to witness the wonders of wave phenomena!

Wave Wonders Unveiled

This simple setup allows us to explore the fascinating world of wave behavior, including:

Reflection: Like a ball bouncing off a wall, waves bounce off a barrier in the tank. The angle at which they hit the barrier (angle of incidence) is mirrored in their rebound (angle of reflection). It's like a perfectly choreographed dance move!
Refraction: Imagine waves approaching a beach. As they move from deeper to shallower water, they slow down and bend, changing direction. This bending, called refraction, can be recreated in our ripple tank by introducing a piece of glass or acrylic, simulating a change in depth.
Diffraction: Ever noticed how sound travels around corners? This is diffraction, the spreading of waves as they pass through a narrow opening or around an obstacle. In our ripple tank, a barrier with a small opening creates a mesmerizing display of wave spreading.
Interference: When two waves meet, they can either amplify each other (constructive interference) or cancel each other out (destructive interference). It's like a wave duet, creating beautiful patterns of peaks and troughs.

Real-World Ripples

The ripple tank isn't just a toy; it's a powerful tool that helps us understand the waves we encounter every day.

Mirrors: These everyday objects use reflection to bounce light back at us, allowing us to see our own reflection.
Lenses: From eyeglasses to cameras, lenses use refraction to bend light and create focused images.
Prisms: These magical objects use refraction to split white light into its rainbow of colors.
Radio Waves: Diffraction allows radio waves to bend around obstacles, ensuring we can enjoy our favorite tunes even when driving through a tunnel.
The ripple tank is a window into the fascinating world of waves, offering a hands-on experience that brings physics to life. So, dive in and explore the magic of wave motion!

Reflection of light

The Law of Reflection: A Mirror to the Universe

Imagine a mischievous sunbeam, dancing through the air, eager to explore the world. It reaches a still pond and, with a playful wink, bounces off the surface, continuing its journey across the meadow. This delightful dance of light is governed by the Law of Reflection, a fundamental principle of optics that reveals the elegant symmetry of light's interaction with the world.

Unveiling the Secrets of Reflection

Before we dive into the depths of this fascinating law, let's familiarize ourselves with the key players in this optical ballet:

The Normal: Picture a tiny flagpole standing perpendicular to a mirror's surface, marking the point where our curious sunbeam strikes. This is the "normal," a reference line that helps us measure the angles of this luminous interaction.
Angle of Incidence: This is the angle between our incoming sunbeam (the incident ray) and the "normal" flagpole.
Angle of Reflection: As our sunbeam bounces off the mirror (now the reflected ray), it forms another angle with the "normal." This is the angle of reflection.
The Law of Reflection, in its elegant simplicity, states that the angle of incidence is always equal to the angle of reflection. It's a cosmic mirror, ensuring that light bounces off a surface at the same angle it arrives.

A Playful Experiment

Grab a flashlight, a mirror, and a protractor, and let's witness this law in action!

Setting the Stage: In a dimly lit room, place your mirror upright on a piece of paper and trace its base. This line represents the reflecting surface.

Raising the Flagpole: Using your protractor, draw a line perpendicular to the mirror's surface at its midpoint. This is our "normal."
Shine the Light: Aim your flashlight at the mirror and observe the reflected beam.
Tracing the Dance: Trace the paths of both the incident and reflected rays on the paper.
Measuring the Angles: With your protractor, measure the angle of incidence and the angle of reflection.
No matter how you change the angle of the flashlight, you'll find that the angle of incidence always matches the angle of reflection. It's a beautiful demonstration of this fundamental law.

Beyond the Mirror: Types of Reflection

While the Law of Reflection holds true universally, the nature of reflection can vary depending on the surface it encounters.

Specular Reflection: When light bounces off a smooth surface like a mirror or a calm lake, it creates a clear, sharp image. This is specular reflection, where the reflected rays remain parallel, like a well-trained marching band.
Diffuse Reflection: In contrast, when light encounters a rough surface like a wall or a piece of paper, the reflected rays scatter in various directions. This is diffuse reflection, resulting in a softer, more diffused image.

Reflection in Our World

The Law of Reflection is not just a theoretical concept; it plays a vital role in our everyday lives and in countless technologies.

Mirrors: From the bathroom mirror that helps us get ready in the morning to the sophisticated mirrors used in telescopes that unveil the mysteries of the universe, mirrors rely on the Law of Reflection.
Optical Instruments: Cameras, telescopes, and microscopes all use mirrors and lenses to manipulate light and create images, guided by the principles of reflection.
Fiber Optics: Fiber optic cables, the backbone of modern communication, transmit data as pulses of light that bounce along the

cable's inner walls through total internal reflection, a phenomenon based on the Law of Reflection.

Beyond the Basics

The Law of Reflection is a gateway to a deeper understanding of light and its interactions. It leads us to explore concepts like refraction, where light bends as it passes from one medium to another, and polarization, where light waves oscillate in specific directions.

Reflection in Action: Case Studies

The Hubble Space Telescope: This iconic telescope, orbiting high above Earth, captures breathtaking images of distant galaxies, relying on precisely shaped mirrors that adhere to the Law of Reflection.
Solar Power: Solar panels, harnessing the sun's energy, utilize anti-reflective coatings to minimize reflection and maximize light absorption, showcasing the practical applications of reflection management.
Animal Vision: Many nocturnal animals, like cats, possess a special layer in their eyes called the tapetum lucidum, which reflects light back through the retina, enhancing their night vision.
The Law of Reflection, in its simplicity and elegance, is a testament to the underlying order of the universe. It's a principle that not only governs the behavior of light but also illuminates our understanding of the world around us.

🪞 ✨ Mirror, Mirror, on the Wall, Who's the Most Mind-Bending of Them All? ✨ 🪞

Forget everything you thought you knew about those shiny surfaces hanging on your wall. We're diving headfirst into the mesmerizing world of plane mirrors, where reality bends, images flip, and the laws of physics get a playful twist.

1. Adventures in Wonderland: A Mirror Experiment Gone Rogue

Who needs textbooks when you can become a mad scientist in your own right? Grab your lab coat (or your pajamas, no judgment here) and get ready for a mind-blowing experiment that will reveal the secrets hidden within the reflective depths.

Your Quest Items:

A Mirror of Mystery: (At least 15 cm x 15 cm, the bigger the better for dramatic effect!)
A Stand-Up Guy: A vertical stand with a clamp to hold your mirror steady.
Doppelganger Duo: Two identical objects (candles, funky figurines, or even your own mismatched socks!).
Ruler of the Realm: A ruler to measure the mysteries of the mirror world.
Angle Master: A protractor to unveil the secrets of reflection.
Blank Canvas: A sheet of white paper to capture your groundbreaking discoveries.
Pencil of Power: A pencil to mark your findings.

Embark on Your Experiment:

Set the Stage: Lay your white paper flat, like a pristine snowfield ready for your scientific footprints. Clamp your mirror to the stand, ensuring its reflective side faces the paper, ready to reveal its secrets.

Summon the Object: Place one of your objects (let's call it Object A, because scientists love labels) a short distance in front of the mirror. Mark its position on the paper with your pencil, like marking a treasure map.

Conjure the Illusion: Now, position the second object (Object B, its mysterious twin) behind the mirror. Move it around until it seems to merge with Object A's reflection, creating a mind-bending illusion. Mark this spot on your paper—you've found the image's secret hideout!

Measure the Magic: With your trusty ruler, measure the distance between Object A and the mirror, and then between Object B and the mirror. Prepare to be amazed—they're the same! It's like the mirror has a secret portal that keeps things perfectly balanced.

Unmask the Illusion: Take a closer look at the image. It's like Object A, but with a twist:

Size Matters Not: The image is the same size as the object, no shrinking or growing allowed in this mirror world.
Upright, But Twisted: The image stands tall, but hold on—it's flipped! Left and right have switched places, like a mischievous mirror prank.
Ghostly Apparition: The image is a phantom, a trick of the light. You can't capture it on a screen, it only exists in the realm of reflections.

Decode the Enigma:

This experiment unveils the secrets of plane mirrors:

Mirror Match: The image is the same size as the object, a perfect match.
Equal Distance: The image hides behind the mirror at the same distance as the object stands before it.
Virtual Reality: The image is a virtual illusion, a meeting point of light rays, not a tangible thing.
Lateral Inversion: The image is flipped horizontally, like a backwards world.

Level Up Your Experiment:

Ready to go deeper down the rabbit hole? Try these mind-bending challenges:

Angle of Reflection: Use your protractor to measure the angle of the incoming light ray and the angle of the reflected ray. They're always equal! It's like the mirror has a strict rulebook for light.
Mirror Maze: Place two mirrors at an angle, and watch the reflections multiply! The number of images depends on the angle, creating a mesmerizing hall of mirrors.

2. The Law of Reflection: The Mirror's Rulebook

The law of reflection is the ultimate rulebook for light bouncing off a surface. It's a simple but powerful rule:

"The angle of incidence (the angle at which light hits the mirror) is equal to the angle of reflection (the angle at which light bounces off)."

This rule explains why mirrors work the way they do, from simple reflections to complex optical illusions.

Unlocking the Secrets of the Law:

Incident Ray: The light ray striking the mirror, like a tiny explorer venturing into the unknown.
Reflected Ray: The light ray bouncing off the mirror, carrying secrets of the reflected world.
Normal: An imaginary line perpendicular to the mirror's surface, like a strict border guard.
Angle of Incidence: The angle between the incident ray and the normal, the measure of the light's approach.
Angle of Reflection: The angle between the reflected ray and the normal, the measure of the light's departure.

The Law in Action:

The law of reflection isn't just some abstract concept—it's the driving force behind a ton of cool stuff:

Plane Mirrors: It explains why your reflection looks the way it does—virtual, laterally inverted, and perfectly positioned.
Periscopes: These sneaky devices use mirrors to let you see around corners, like a submarine peeking above the waves.
Sextants: Sailors use these tools to navigate by the stars, using mirrors and angles to find their way across the vast ocean.
Optical Squares: Surveyors use these gadgets to measure distances and create accurate maps, all thanks to the magic of reflection.
Reflectors: Car headlights, flashlights, and telescopes use reflectors to focus light and make it shine brighter or travel farther.
Fiber Optics: The internet wouldn't exist without fiber optics, which use reflection to transmit information as light pulses through tiny glass fibers.
Kaleidoscopes: These mesmerizing toys create endless patterns of color and light, all thanks to the law of reflection.

Mirror, Mirror, in the Tech:

Hubble Space Telescope: This incredible telescope uses giant mirrors to capture images of distant galaxies, revealing the wonders of the universe.
Solar Power Plants: Massive arrays of mirrors focus sunlight to generate clean energy, harnessing the power of reflection for a sustainable future.
Laser Scanners: From scanning groceries at the checkout to creating detailed 3D models, laser scanners rely on the precise reflection of laser beams.

The End... or is it?

The world of plane mirrors is a portal to endless possibilities. By experimenting, observing, and understanding the law of reflection, we can continue to unlock the secrets of light and create amazing new technologies. So go forth, explore, and reflect—the world is your mirror! 🪞 ✨

Refraction of light

Imagine Light as a Mischievous Child:

Picture light as a playful child, always eager to explore and take the quickest route. When it travels through the same material, like air, it zips along in straight lines, just like a child running across an open field. But when it reaches a new material, like water, it's like the child suddenly encountering a muddy patch. The child slows down and changes direction to avoid getting stuck, and so does light! This bending of light's path is what we call refraction.

The "Stickiness" of Stuff:

Think of each material as having a different "stickiness" for light. A denser material, like glass, is "stickier" than air. When light enters a "stickier" material, it slows down and bends towards an imaginary line called the normal – this line is like an invisible guidepost pointing straight out from the surface. The opposite happens when light goes from a "sticky" material to a less "sticky" one; it speeds up and bends away from the normal.

Snell's Law: The Secret Code of Light:

This "stickiness" of a material is measured by something called the refractive index. A clever scientist named Snell figured out a way to predict exactly how much light will bend, and he wrote it down in a secret code called Snell's Law. It's like a magic formula that tells you the angle at which light will bend when it goes from one material to another.

Light's Magical Tricks:

Refraction isn't just some boring science concept; it's the reason we see amazing things every day!

Rainbows: Imagine tiny raindrops acting like prisms, splitting sunlight into a dazzling spectrum of colors. That's refraction in action!
Sparkling Diamonds: The way a diamond sparkles are due to light bouncing around inside it and being refracted many times before it finally escapes, creating a dazzling display of brilliance.
Mirages: Have you ever seen a shimmering "puddle" on a hot road that disappears as you get closer? That's a mirage, a trick of the light caused by refraction in the hot air near the ground.
Fiber Optic Cables: These thin strands of glass use refraction to trap light inside, allowing it to travel for miles and carrying information with it – that's how the internet works!

The World Through Refraction's Lens:

Refraction isn't just about pretty effects; it's also incredibly useful. Eyeglasses, microscopes, telescopes, and cameras all use lenses that bend light to help us see things better, whether it's the tiny details of a cell or the vastness of the universe.

Keep Exploring!

This is just the beginning of your journey into the fascinating world of refraction. There's so much more to discover, from the way light splits into different colors to the creation of invisibility cloaks! So, keep asking questions, keep exploring, and let the magic of light guide you.

Refraction of Light Through Transparent Blocks and Total Internal Reflection

1. Experiment Demonstrating Refraction of Light Through Transparent Blocks

Objective:

To observe and understand how light bends as it passes through different transparent blocks.

Materials:

Ray box or laser pointer
Transparent blocks of various shapes (rectangular, triangular, semi-circular)
White paper
Protractor
Ruler
Pencil

Procedure:

Set the Stage: Lay a white sheet of paper on a flat surface. Position the ray box to cast a narrow beam of light onto the paper.

Trace the Incident Ray: Trace the path of the light ray on the paper. This is the incident ray.
Introduce the Block: Place one of the transparent blocks on the paper, intercepting the light ray.
Trace the Refracted Ray: Observe the light ray changing direction as it enters and exits the block. Trace the path of the ray through the block and as it exits. This is the refracted ray.
Mark Angles: Mark the point where the light ray enters the block. Draw a normal line (perpendicular to the block's surface) at this point.

Measure and label the angle of incidence (the angle between the incident ray and the normal) and the angle of refraction (the angle between the refracted ray and the normal).
Repeat with Different Shapes: Repeat steps 3-5 with different shaped blocks. Observe how the light ray bends differently depending on the shape of the block and the angle of incidence.
Vary the Angle of Incidence: For each block, repeat the experiment with different angles of incidence. Notice how the angle of refraction changes with the angle of incidence.

Observations:

The light ray bends towards the normal as it enters the block from air (a less dense medium to a denser medium).
The light ray bends away from the normal as it exits the block into air (a denser medium to a less dense medium).
The degree of bending depends on the shape of the block and the angle of incidence.
For a given block, the angle of refraction increases as the angle of incidence increases.

Analysis:

This experiment demonstrates the phenomenon of refraction, where light changes direction as it passes from one medium to another. The degree of bending depends on the refractive indices of the two media and the angle of incidence. Snell's Law governs this relationship:

n₁sinθ₁ =
n₂sinθ₂

where:

n₁ is the refractive index of the first medium
θ₁ is the angle of incidence
n₂ is the refractive index of the second medium
θ₂ is the angle of refraction

Additional Notes:

You can use a laser pointer instead of a ray box for a more focused beam of light.
Use a protractor with a clear base for more accurate angle measurements.
Ensure the room is dimly lit to enhance the visibility of the light ray.
Repeat the experiment multiple times for each block and angle of incidence to ensure reliability.
You can extend this experiment by investigating how the color of light affects refraction.

2. Critical Angle and Total Internal Reflection

Critical Angle:

When light travels from a denser medium to a less dense medium (e.g., from glass to air), it bends away from the normal. As the angle of incidence increases, the angle of refraction also increases. At a certain angle of incidence, the angle of refraction becomes 90 degrees. This specific angle of incidence is known as the critical angle.

Total Internal Reflection:

If the angle of incidence is increased beyond the critical angle, the light ray no longer refracts but is entirely reflected back into the denser medium. This phenomenon is called total internal reflection.

Conditions for Total Internal Reflection:

Two conditions must be met for total internal reflection to occur:

Light must travel from a denser medium to a less dense medium.
The angle of incidence must be greater than the critical angle.
Equation:

The relationship between the critical angle (c) and the refractive index (n) of the denser medium is given by the equation:

n = 1/sin c

where:

n is the refractive index of the denser medium
c is the critical angle
Derivation:

This equation can be derived from Snell's Law. When the angle of incidence is equal to the critical angle (c), the angle of refraction is 90 degrees. Substituting these values into Snell's Law:

n₁sin c = n₂sin 90°

Since sin 90° = 1 and considering n₁ as the refractive index of the denser medium (n) and n₂ as the refractive index of the less dense medium (usually air, with n $\approx$ 1), the equation simplifies to:

n sin c = 1

Therefore:

n = 1/sin c

Examples:

Diamond: Diamonds have a high refractive index (n $\approx$ 2.42), resulting in a small critical angle (c $\approx$ 24.4°). This leads to significant internal reflection, making diamonds sparkle.
Optical Fibers: Optical fibers utilize total internal reflection to transmit data over long distances with minimal loss. Light signals are trapped within the fiber by repeated total internal reflections, allowing them to travel along the fiber's length.

Case Studies:

Endoscopy: This medical procedure uses optical fibers to examine the internal organs of the body. Light is transmitted through one set of

fibers to illuminate the internal organs, while another set of fibers carries the reflected light back to the viewing device.
Prism Binoculars: Binoculars use prisms to fold the optical path, making them more compact. The prisms utilize total internal reflection to reflect the light, ensuring that the image remains upright and correctly oriented.
Fiber Optic Communication: This technology uses optical fibers to transmit data as light pulses. Total internal reflection keeps the light signals confined within the fiber, allowing for high-speed, long-distance communication.

Detailed Explanation:

Total internal reflection is a fascinating phenomenon with numerous applications in various fields. It occurs due to the relationship between the angle of incidence and the refractive indices of the two media. When light travels from a denser medium to a less dense medium at an angle greater than the critical angle, it cannot refract out of the denser medium and is instead reflected back into it. This principle is exploited in various optical devices, such as optical fibers, prisms, and even in natural phenomena like mirages. The equation $n = 1/\sin c$ provides a fundamental relationship between the refractive index of a material and its critical angle, allowing us to understand and predict the behavior of light in different media.

This comprehensive discussion provides a deeper understanding of refraction, critical angle, total internal reflection, and the equation $n = 1/\sin c$, highlighting their significance in both theoretical and practical contexts.

Unveiling the Magic of Light: A Journey into Internal Reflection

Imagine light, not as a simple ray, but as a mischievous sprite, dancing through the air and playing tricks on our eyes. Now, picture it encountering a shimmering surface, like a pool of water or a crystal prism. Instead of passing straight through, the light sprite decides to bounce back, trapped within the denser medium. This, my friend, is the captivating phenomenon of internal reflection.

The Glass Block Adventure

Let's embark on a little experiment, shall we? Grab a semi-circular glass block, a ray box, and a sheet of paper. As you shine the light into the block, watch as the sprite bends its path, almost as if it's reluctant to leave the cozy confines of the glass. Increase the angle, and the sprite bends even further, until it reaches a point where it can no longer escape, gliding along the surface like a graceful skater. This is the critical angle, the threshold beyond which the sprite is forever bound to the glass.

The Water Tank Illusion

Now, let's dive into a water tank, armed with a laser pointer. As the beam enters the water, it seems to bend, creating an illusion of a broken stick. But the real magic happens when you increase the angle. Suddenly, the beam bounces back, refusing to emerge from the water's surface. It's as if the water has cast a spell, trapping the light within its depths.

Total Internal Reflection: A Light Trap

But wait, there's more! Let's introduce a prism, a master of light manipulation. As the light enters the prism, it bends towards the normal, but as it tries to exit, it's met with an invisible barrier. The angle is too steep, and the light sprite is completely reflected back inside, bouncing around like a trapped firefly. This is total internal reflection, a phenomenon that has revolutionized the way we communicate.

Optical Fibers: Whispers of Light

Think of an optical fiber as a tiny tunnel, guiding light along its path, even when it bends and curves. This is possible thanks to total internal reflection, where the light bounces off the walls of the fiber, never escaping its confines. It's like whispering secrets into a long tube, the message carried faithfully by the light sprite.

The Wonders of Optical Communication

Optical fibers have transformed the way we connect, enabling lightning-fast internet, crystal-clear cable TV, and seamless mobile networks. They're the invisible threads that weave our digital world together, carrying vast amounts of data across continents and oceans.

A World of Possibilities

From long-distance communication to high-speed internet, optical fibers have become the backbone of our modern world. They're the silent heroes of the digital age, enabling us to connect, communicate, and explore like never before.

So, the next time you see a shimmering reflection or use your high-speed internet, remember the magic of internal reflection, the captivating dance of light that has transformed our world.

Thin lenses

Lenses: The Magnificent Light Benders

Imagine light as a mischievous stream of tiny particles, always eager to explore the world. Now, picture lenses as magical gateways that can guide these particles, bending their paths and shaping their journeys.

Converging Lenses: The Light Gatherers

Converging lenses, also known as convex lenses, are like friendly giants with a plump middle. When light encounters these lenses, it's as if they have a welcoming embrace, gently bending the light inwards towards a single point. This point, where the light rays converge, is called the focal point, a place where the light's energy is concentrated.

Think of a magnifying glass, a tool that can make tiny objects appear larger than life. This magic is possible because of a converging lens, which gathers the scattered light from an object and focuses it onto our eyes, creating an enlarged image.

Diverging Lenses: The Light Spreaders

Diverging lenses, or concave lenses, are like playful elves with a slender middle. When light encounters these lenses, it's as if they have a mischievous touch, playfully spreading the light outwards. The light rays appear to diverge from a point called the virtual focus, creating an illusion of a light source behind the lens.

These lenses are the secret behind eyeglasses that help people with nearsightedness see clearly. By spreading the incoming light, diverging lenses ensure that the image is formed correctly on the retina, allowing the wearer to see the world in sharp focus.

Focal Length: The Measure of a Lens's Power

The focal length of a lens is like its unique fingerprint, a measure of its ability to bend light. A shorter focal length means a stronger lens, one that can bend light more dramatically. It's like comparing a gentle curve in a river to a sharp bend in a mountain road.

Principal Axis: The Guiding Line

The principal axis is like an invisible ruler that runs through the center of the lens, helping us understand how the light behaves. It's like a central landmark that guides the light on its journey through the lens.

Applications: Lenses in Action

Lenses are the unsung heroes of countless optical devices, from cameras that capture precious moments to telescopes that reveal the wonders of the universe. They are the eyes of microscopes that unveil the hidden world of tiny organisms and the heart of projectors that bring movies to life on the big screen.

Beyond the Basics

The world of lenses is full of fascinating phenomena, like lens aberrations that can distort images and advanced lens systems that can correct these distortions. It's a world where science and magic intertwine, where light and glass dance together to create amazing visual experiences.

Conclusion

Lenses are not just pieces of glass; they are the architects of light, shaping its path and revealing the wonders of the world around us. They are the tools that allow us to see the big and the small, the near and the far, the visible and the invisible.

The Lens That Bends Light and Our Perception: A Journey into Converging Lenses

Imagine a humble magnifying glass, the kind Sherlock Holmes might have used to scrutinize a clue. This simple tool, a converging lens, holds the power to bend light and alter our perception of the world. Let's embark on a journey to uncover the magic of converging lenses, exploring how they form images and magnify the unseen.

Converging Lenses: Where Light Converges

Picture a lens that bulges in the middle, like a lentil or a drop of water. This is a converging lens, also known as a convex lens. As light rays pass through this lens, they bend inward, converging towards a single point, aptly named the focal point. The distance between the lens' center and this focal point is its focal length, a key factor in determining the lens's magnifying power.

These lenses are ubiquitous in our lives, from the eyeglasses that correct vision to the cameras that capture memories. They even empower scientific instruments like microscopes and telescopes, revealing the intricate details of tiny organisms and the vastness of distant galaxies.

The Dance of Light and Images

The position of an object relative to the focal point determines the type of image formed by a converging lens. Let's delve into the two main scenarios:

1. Real Images: Projections of Reality

When an object is placed beyond the focal point, a real image is born. This image is inverted, meaning it's upside down, and can be either larger or smaller than the object depending on its distance from the lens. The magic of real images lies in their ability to be projected onto a screen, just like a movie projector casting a film onto a screen for all to see.

2. Virtual Images: Illusions of Light

When an object ventures closer, nestled between the focal point and the lens, a virtual image emerges. This image is upright and always magnified, appearing to reside behind the lens. Unlike its real counterpart, a virtual image cannot be projected onto a screen; it's an illusion created by the apparent divergence of light rays. Think of a magnifying glass creating a larger-than-life image of a tiny insect.

Linear Magnification: Measuring the Illusion

Linear magnification, denoted by 'M', quantifies how much larger or smaller an image is compared to the original object. It's the ratio of the image height to the object height. A magnification greater than 1 signifies an enlarged image, while a value between 0 and 1 indicates a diminished image.

Case Studies: Lenses in Action

Let's witness the power of converging lenses in real-world scenarios:

The Camera Lens: A camera captures moments by forming a real, inverted image on its sensor. By adjusting the lens's distance from the sensor, the photographer fine-tunes the focus, while the focal length determines the magnification, allowing for wide-angle or zoomed-in shots.

The Human Eye: Our eyes house a remarkable converging lens, the crystalline lens, which focuses light onto the retina, our internal screen. The eye's ability to change the lens's shape, known as accommodation, allows us to focus on objects near and far.

The Telescope: Telescopes, humanity's windows to the cosmos, employ two converging lenses. The objective lens gathers light from distant celestial bodies, while the eyepiece magnifies the image, bringing the wonders of the universe closer to our eyes.

Conclusion: The Enduring Magic of Lenses

Converging lenses are more than just pieces of glass; they are instruments of perception, bending light to reveal the unseen and magnify the minute. From capturing memories to exploring the cosmos, these lenses play a vital role in our lives. By understanding their principles, we gain a deeper appreciation for the magic they bring to our world.

The Humble Magnifying Glass: A Portal to Hidden Worlds

Imagine holding a tiny universe in your hand, a lens that unveils the intricate details of the world around us. That's the magic of a magnifying glass, a simple convex lens that bends light to create a larger, virtual image. It's a tool that has fascinated curious minds for centuries, from children examining ants to scientists exploring the microscopic realm.

A Journey Through the Lens

When you peer through a magnifying glass, you're not just seeing things bigger; you're entering a world of hidden wonders. The delicate veins of a leaf, the shimmering facets of a gemstone, the tiny gears of a watch – all become magnified, revealing their intricate beauty.

The power of a magnifying glass lies in its focal length, the distance from the lens to the point where light rays converge. A shorter focal length means greater magnification, bringing the unseen into focus.

More Than Just Magnification

But the magnifying glass is more than just a tool for seeing things bigger. It's a symbol of curiosity, a key to unlocking the secrets of the world around us. It's a reminder that there's always more to discover, if we only take the time to look closer.

The Eye: A Natural Magnifying Glass

Our own eyes are marvels of nature, with lenses that focus light onto the retina, creating the images we see. But sometimes, our eyes need a little help. That's where corrective lenses come in, bending light to compensate for nearsightedness or farsightedness.

A World of Possibilities

From reading fine print to inspecting intricate objects, the magnifying glass has countless applications. It's a tool for exploration, a window into the hidden wonders of the world around us. So next time you pick up a magnifying glass, remember that you're holding more than just a lens – you're holding a portal to a universe of discovery.

Dispersion of light

A Prism's Rainbow: Unlocking the Secrets of Light

Imagine a beam of sunlight, seemingly simple and pure, suddenly exploding into a riot of colors as it passes through a clear, triangular prism. This magical transformation, known as dispersion, is a testament to the complex nature of light itself.

Light's Bending Journey: Refraction

Light, like a traveler, changes its pace when it moves from one medium to another. This change in speed causes it to bend, a phenomenon called refraction. Think of it like a car hitting a patch of mud – it slows down and veers off course. Similarly, when light enters a denser medium like glass, it slows down and bends.

But here's the twist: not all colors of light bend the same way. Shorter wavelengths, like the energetic violet and blue, take a sharper turn, while longer wavelengths, like the mellow red and orange, take a gentler curve. It's like a group of friends running through a maze – the quickest one's zip around corners, while the slower ones take a wider path.

Prism: The Light Splitter

A prism, with its flat, polished surfaces, acts like a cunning magician, bending light not once, but twice. As light enters and exits the prism, it's like going through two sets of bends in the road. This double bending, combined with the different bending angles of each color, causes the white light to fan out into its individual components, creating a dazzling spectrum.

The Spectrum: A Symphony of Colors

The spectrum, a mesmerizing band of colors, unfolds like a rainbow. Traditionally, we divide it into seven distinct hues:

Red: The longest wavelength, like a leisurely stroll, bends the least.
Orange:
Yellow:
Green:
Blue:
Indigo:
Violet: The shortest wavelength, like a quick sprint, bends the most.

Why the Split?

The secret to this colorful separation lies in the prism's refractive index, a measure of how much it bends light. This index varies slightly for different wavelengths, meaning each color experiences a different degree of bending. It's like a group of people walking on a slanted surface – those with shorter strides will veer off more than those with longer strides.

Factors Influencing the Spread

The extent to which light disperses depends on a few key factors:

Prism Material: Different materials have different refractive indices, affecting the degree of bending. It's like choosing between a winding mountain road and a straight highway – the path you choose determines how much you deviate from your original direction.

Prism Angle: The angle between the prism's faces also plays a role. A steeper angle leads to a greater separation of colors.
Wavelength of Light: As we've seen, shorter wavelengths bend more than longer ones.

Dispersion in Action

Dispersion isn't just a pretty spectacle; it has real-world applications:

Spectroscopy: Scientists use dispersion to analyze the light emitted or absorbed by substances, revealing their chemical composition. It's like identifying a person by their unique fingerprint.
Optical Instruments: Prisms are essential components in binoculars, telescopes, and cameras, helping to focus and manipulate light.
Color Displays: Liquid crystal displays, found in our phones and TVs, rely on the principles of dispersion to create vibrant images.

Beyond the Visible

While we marvel at the rainbow of colors we can see, dispersion also occurs for invisible forms of light, like ultraviolet and infrared. It's a reminder that there's more to the world than meets the eye.

The Spectrum's Colorful Symphony

The visible spectrum is just a small slice of the electromagnetic spectrum, the vast range of energy that travels in waves. Each color in the spectrum vibrates at a different frequency, like musical notes forming a harmonious chord.

Frequency and Wavelength: A Balancing Act

Frequency, the number of waves passing a point in a second, is inversely related to wavelength, the distance between wave peaks. High frequency means short wavelength, and vice versa. It's like the relationship between steps taken and distance covered – the smaller your steps, the more you need to take to cover the same distance.

Nature's Colorful Palette

Dispersion paints our world with vibrant hues:

Rainbows: These majestic arcs of color are born from sunlight dispersed by raindrops.
Aurora Borealis: The mesmerizing dance of colors in the night sky is a result of charged particles from the sun interacting with Earth's atmosphere.
Butterfly Wings: The iridescent colors of butterfly wings often arise from the interference and diffraction of light.

Conclusion

The dispersion of light by a prism is a captivating phenomenon that reveals the hidden complexity of light. By understanding refraction and the spectrum, we can appreciate the beauty and wonder of the world around us.

Electromagnetic spectrum

The Electromagnetic Spectrum: A Symphony of Light and Waves

Imagine a grand orchestra, with instruments ranging from the deepest bass to the highest piccolo. Each instrument plays its part, contributing to a rich and complex symphony. The electromagnetic (EM) spectrum is much like this orchestra, a vast and harmonious ensemble of energy waves, each with its unique frequency and wavelength.

A Journey Through the Spectrum

Let's embark on a journey through this fascinating spectrum, exploring its diverse regions and their remarkable applications.

Radio Waves: The Long-Distance Communicators

These gentle giants of the EM spectrum have the longest wavelengths and lowest energy. They're the backbone of our communication networks, carrying radio and television broadcasts, enabling Wi-Fi and Bluetooth connections, and guiding GPS satellites.

Microwaves: The Heat-Bringers

Microwaves are shorter than radio waves and pack a bit more energy. They're the chefs of the EM spectrum, cooking our food in microwave ovens. They also play a vital role in satellite communications and radar systems, helping us track weather patterns and navigate aircraft.

Infrared Radiation: The Invisible Warmth

Infrared radiation is often associated with heat. We can't see it, but we can feel it as warmth from the sun or a cozy fire. It's also used in night vision goggles, remote controls, and thermal imaging cameras, allowing us to see in the dark and detect heat signatures.

Visible Light: The Rainbow's Embrace

This is the only part of the EM spectrum that our eyes can perceive, a dazzling spectacle of colors ranging from red to violet. Visible light illuminates our world, allowing us to see the beauty around us and capture it in photographs.

Ultraviolet Radiation: The Sun's Energetic Kiss

Ultraviolet radiation is more energetic than visible light and can cause sunburn if we're not careful. But it also has beneficial effects, such as stimulating vitamin D production in our skin. UV light is used in sterilization, fluorescence, and photolithography, shaping the microscopic world of electronics.

X-rays: The Bone-Piercing Detectives

X-rays have very short wavelengths and high energy, allowing them to penetrate soft tissues but not dense materials like bone. They're invaluable in medical imaging, security screening, and materials analysis, revealing hidden structures and potential threats.

Gamma Rays: The Cosmic Powerhouses

Gamma rays are the most energetic members of the EM spectrum, born from violent cosmic events like supernovae and nuclear explosions. They're used in cancer treatment, sterilization, and industrial imaging, harnessing their immense power for both healing and exploration.

The Speed of Light: A Universal Constant

All electromagnetic waves, regardless of their frequency or wavelength, travel at the same astonishing speed in a vacuum: approximately 186,282 miles per second. This universal speed limit, denoted by the symbol 'c', is a cornerstone of modern physics.

The EM Spectrum in Action

The EM spectrum is woven into the fabric of our lives, from the mundane to the extraordinary. It's the foundation of our communication networks, the source of our medical imaging technologies, and the key to unlocking the secrets of the universe.

Emerging Trends and Future Frontiers

The EM spectrum continues to surprise and inspire us with new discoveries and applications. Terahertz technology is opening up exciting possibilities in security screening, medical imaging, and high-speed communication. 5G networks are pushing the boundaries of wireless communication, while quantum communication promises unparalleled security. Advances in medical imaging are leading to more precise and less invasive diagnostic tools, and light-based computing may revolutionize the future of technology.

The EM Spectrum: A Tapestry of Wonder

The electromagnetic spectrum is a testament to the beauty and complexity of our universe. It's a symphony of light and waves, a tapestry of energy and information, and a source of endless fascination and exploration. As we continue to unravel its mysteries, we can expect even more remarkable discoveries and innovations in the years to come.

Unveiling the Invisible: A Journey Through the World of Radio Waves, Microwaves, and Infrared

Imagine a world without music playing on your radio as you commute, without the ability to quickly reheat your lunch in a microwave, or without the convenience of changing channels on your TV from the comfort of your couch. This is a world without radio waves, microwaves, and infrared radiation – three types of electromagnetic waves that play a crucial role in our modern lives.

Let's embark on a journey to uncover the magic behind these invisible forces and explore their diverse applications.

Radio Waves: The Long-Distance Communicators

Radio waves are the marathon runners of the electromagnetic spectrum, capable of traveling vast distances through the air and even solid objects. They are the backbone of our communication systems, bringing us radio and television broadcasts, connecting our mobile phones, and even revealing the secrets of the universe.

Radio and Television: From the crackly broadcasts of early radios to the crystal-clear images of digital television, radio waves have revolutionized how we consume entertainment and information. They allow us to tune into our favorite radio stations, catch up on the latest news, and immerse ourselves in captivating TV shows.

Astronomy: Radio waves emanating from distant stars and galaxies provide a unique window into the cosmos. Radio telescopes, like giant ears listening to the whispers of the universe, capture these faint signals and help astronomers unravel the mysteries of black holes, pulsars, and the Big Bang.

Microwaves: The Multi-talented Waves

Microwaves, with their shorter wavelengths, are the versatile workhorses of the electromagnetic spectrum. They are used in a wide range of applications, from heating our food to transmitting satellite signals and powering our mobile phones.

Satellite Communication: Microwaves are the invisible threads that connect us to satellites orbiting high above the Earth. They enable satellite television broadcasts, GPS navigation, and long-distance communication.

Mobile Phones: Every time you make a call or browse the internet on your mobile phone, you are relying on the power of microwaves. They carry your voice and data between your phone and cell towers, connecting you to the world.

Microwave Ovens: Microwaves have revolutionized cooking, allowing us to quickly heat food by exciting water molecules. They have become an indispensable kitchen appliance, saving us time and effort.

Infrared: The Heat-Carrying Rays

Infrared radiation, invisible to the human eye, is the warmth we feel from the sun or a cozy fire. It is also used in a variety of technologies, from remote controls to thermal imaging cameras.

Remote Controls: Infrared LEDs in remote controls send coded signals to our TVs, stereos, and other devices, allowing us to control them from a distance.

Thermal Imaging: Infrared cameras detect the heat emitted by objects, creating images based on temperature differences. This technology is used in night vision goggles, medical diagnostics, and building inspections.

Heating: Infrared heaters provide efficient and targeted warmth, directly heating objects and people rather than the surrounding air. They are commonly used in outdoor patios, workshops, and even medical therapies.

The Future of Electromagnetic Waves

As technology continues to advance, the applications of radio waves, microwaves, and infrared radiation are expanding. From the development of 5G networks to the use of infrared sensors in self-driving cars, these invisible forces are shaping the future of communication, healthcare, and countless other fields.

So, the next time you listen to the radio, heat up your lunch in the microwave, or change the channel on your TV, take a moment to appreciate the invisible wonders of radio waves, microwaves, and infrared radiation. They are the silent partners in our modern lives, connecting us, informing us, and making our lives easier and more enjoyable.

The Dance of Light: How We See and Capture the World

Imagine light as a mischievous sprite, zipping through the universe and bouncing off everything it touches. This playful dance of light is what allows us to see the world in all its vibrant glory, and it's also the secret ingredient that makes photography possible.

Think of your eye as a tiny, intricate camera. Light tiptoes through your pupil, a tiny doorway into a darkened room. It then twirls through the lens, a magical crystal that bends and focuses the light onto the retina, a canvas at the back of your eye. Here, millions of tiny sensors, like a chorus of excited spectators, burst into action, translating the light into electrical signals that race along the optic nerve to the brain. And voila! Your brain, the master interpreter, transforms these signals into the images you see.

But what about color, you ask? Ah, that's where the magic really happens! Light isn't just a single entity; it's a kaleidoscope of different colors, each with its own unique wavelength. Our eyes have special sensors called cones, each tuned to a different part of this colorful spectrum. When light hits these cones, they sing out their individual tunes, and our brain blends these melodies into the symphony of color we perceive.

Now, let's bring photography into the picture. A camera is like a mechanical eye, capturing the dance of light and freezing it in time. The lens, a glass eye, gathers the light and focuses it onto a sensor, a digital canvas. This sensor, studded with millions of tiny light-sensitive pixels, eagerly drinks in the light and transforms it into a digital image.

But photography isn't just about capturing what we see; it's about shaping light to tell a story. Photographers are like light whisperers, using tools like lenses, filters, and flashes to manipulate light and shadow, creating moods and emotions. They chase the golden hour, when the sun paints the world in warm hues, and they embrace the drama of side lighting, where shadows dance and highlights gleam.

Think of landscape photographers patiently waiting for the perfect light to bathe a mountain range in an ethereal glow. Or portrait photographers skillfully using soft boxes to create a flattering light that gently caresses their subject's features. And sports photographers, armed with powerful strobes, freezing the fleeting moment of an athlete in motion.

But sometimes, our eyes and cameras can falter. Color blindness can mute the vibrancy of the world, while conditions like myopia or cataracts can blur our vision. Yet, even in these challenges, the dance of light continues, reminding us of its power and beauty.

So, the next time you gaze at a breathtaking sunset or marvel at a photograph, remember the incredible journey of light. It's a journey that connects us to the world, allowing us to see, to create, and to share the magic of our visual experience.

Unveiling the Invisible: X-rays and Gamma Rays - Our Allies in Medicine, Security, and Beyond

Imagine being able to peer through solid objects, like a superhero with X-ray vision! While we can't quite do that yet, we have harnessed the power of invisible light – X-rays and gamma rays – to accomplish incredible feats in various fields.

These powerful forms of electromagnetic radiation, with their extraordinary ability to penetrate materials, have revolutionized medicine, security, and industry. Let's delve into their fascinating world and uncover their diverse applications.

X-rays: Illuminating the Hidden

1. Medical Imaging: A Window into the Human Body

X-rays are like magic wands for doctors, allowing them to see inside our bodies without resorting to surgery. They act like a spotlight, shining through our tissues and creating a shadow play of our internal structures. Denser tissues, like bones, cast darker shadows, while softer tissues, like organs, appear lighter. This contrast helps doctors diagnose a myriad of conditions, from fractures to tumors.

Radiography: The workhorse of medical imaging, used for quick checks on broken bones, lung infections, and more.
Mammography: A specialized technique for detecting breast cancer, empowering women to take control of their health.
Computed Tomography (CT): Imagine taking multiple X-ray snapshots from different angles and stitching them together to create a 3D model of the body. That's CT scanning in a nutshell! It provides intricate details, helping doctors pinpoint problems with incredible accuracy.
Fluoroscopy: Like a live X-ray video, fluoroscopy allows doctors to observe movement in real-time, guiding them during surgeries or observing how we swallow.

A Real-Life Hero:

Picture this: A patient rushes into the ER with agonizing abdominal pain. A quick CT scan reveals a ruptured appendix, a life-threatening condition. Thanks to the X-ray's power, surgeons can swiftly intervene, saving the day.

2. Security Scanners: Guardians of Safety

X-ray scanners are the silent sentinels at airports and other secure locations, ensuring our safety by detecting hidden threats. Just like in medical imaging, they exploit the fact that different materials absorb X-rays differently. This allows security personnel to identify concealed weapons, explosives, or other prohibited items, keeping us safe from harm.

Baggage Scanners: These trusty machines scan our luggage, ensuring we're not carrying anything dangerous onboard.
Full-Body Scanners: Using advanced technologies like backscatter X-ray or millimeter waves, these scanners can detect objects hidden under clothing, all while minimizing radiation exposure.
Averted Danger:

Imagine a passenger attempting to board a plane with a concealed weapon. Thanks to the vigilant X-ray scanner, the threat is detected, preventing a potential tragedy.

3. Cancer Treatment: A Double-Edged Sword

While high doses of X-rays can be harmful, they can also be harnessed to fight cancer. In radiation therapy, X-rays are used to precisely target and destroy cancer cells, offering hope to countless patients.

External Beam Radiation Therapy: X-rays are delivered from outside the body, like a focused beam of energy, targeting the tumor with pinpoint accuracy.
Brachytherapy: Radioactive sources are placed inside or near the tumor, delivering a concentrated dose of radiation directly to the enemy.

A Beacon of Hope:

A patient with lung cancer undergoes radiation therapy, and the tumor shrinks significantly, improving their quality of life and offering a chance at recovery.

4. Engineering Applications: Ensuring Quality and Safety

X-rays are also invaluable in engineering, where they are used to inspect the internal structure of materials and components without causing any damage. This is crucial for ensuring the safety and reliability of critical structures like bridges, pipelines, and aircraft.

Weld Inspection: X-rays can reveal hidden flaws in welds, preventing potential failures in structures we rely on every day.
Casting Inspection: Detecting internal defects in castings ensures the quality and durability of components used in various industries.
Aircraft Inspection: X-rays help identify fatigue cracks and corrosion in aircraft, ensuring flight safety and preventing accidents.

Preventing Disaster:

An X-ray inspection of a pipeline weld reveals a tiny crack. Thanks to this early detection, the crack is repaired, preventing a potential rupture and environmental catastrophe.

Gamma Rays: The Mighty Powerhouse

1. Medical Treatment: A Force for Healing

Gamma rays, with their even greater energy, are also used in cancer treatment and medical sterilization. They can penetrate deep into tissues, targeting tumors with incredible precision.

Radiation Therapy: Similar to X-rays, gamma rays can be used in external beam radiation therapy and brachytherapy to eliminate cancer cells.
Stereotactic Radiosurgery: A highly precise technique that delivers a high dose of gamma rays to a small tumor, minimizing damage to surrounding healthy tissue.
Gamma Knife: A specialized device that uses multiple beams of gamma rays to precisely target brain tumors and other abnormalities, offering a non-invasive alternative to traditional surgery.

A Life-Saving Intervention:

A patient with a brain tumor undergoes Gamma Knife surgery. The procedure successfully destroys the tumor without the need for invasive open surgery, preserving precious brain tissue.

2. Sterilization: Ensuring Safety and Hygiene

Gamma rays are also champions of hygiene, used to sterilize medical equipment and food products. Their penetrating power allows them to kill microorganisms even through packaging, ensuring safety and preventing infections.

Medical Equipment Sterilization: Gamma irradiation sterilizes disposable medical supplies like syringes and surgical gloves, protecting patients from harmful bacteria and viruses.
Food Sterilization: Gamma irradiation can extend the shelf life of food by eliminating bacteria, mold, and insects, while preserving its quality and nutritional value.

Protecting Our Health:

Spices are irradiated with gamma rays to eliminate harmful bacteria like Salmonella, ensuring food safety and preventing foodborne illnesses.

Conclusion: X-rays and Gamma Rays - Our Invisible Allies

X-rays and gamma rays, these invisible forms of light, have become indispensable tools in modern society. They illuminate the hidden, protect our safety, heal our bodies, and ensure the quality of our products. As we continue to explore their potential, we can expect even more groundbreaking applications in the future.

The Unseen Threat: How Electromagnetic Radiation Affects Our Everyday Lives

Imagine a world without your smartphone, your microwave, or even the sun's warm rays. Impossible, right? Yet, these everyday comforts come with a hidden cost: electromagnetic radiation (EMR). While EMR is essential for life, it's a double-edged sword. Like a mischievous sprite, it can be both beneficial and harmful, depending on its form and intensity.

The Burning Touch of Non-Ionizing Radiation

Think of non-ionizing radiation as a mischievous imp. It can't break things directly, but it can cause a ruckus by shaking things up. When this imp encounters your body's molecules, it makes them vibrate, generating heat.

Microwaves: Your trusty kitchen companion uses high-power microwaves to heat your food. But if those waves escape, they can cause burns, especially to sensitive areas like the eyes. Imagine a child, curious about the humming box, peeking inside while it's running. A momentary lapse in supervision could lead to painful injuries.

RF Diathermy: This medical procedure uses radiofrequency waves to generate heat within body tissues for therapeutic purposes. But like a mischievous fire spirit, if not controlled properly, it can cause burns instead of healing.

Radar Systems: These powerful systems, used for navigation and weather forecasting, emit intense radio waves. For technicians working near these systems, it's like being exposed to an invisible bonfire, requiring strict safety measures to prevent burns.

The Silent Scourge of Ionizing Radiation

Ionizing radiation is like a stealthy ninja, capable of causing damage at the cellular level. It can knock electrons off atoms, creating ions that disrupt DNA and other vital molecules.

Ultraviolet Radiation: The sun's warm embrace also carries harmful UV rays. Imagine these rays as tiny archers, their arrows capable of piercing your skin and damaging your DNA. This can lead to skin cancer, including the deadly melanoma.

X-rays and Gamma Rays: These high-energy forms of radiation are like powerful swords, capable of slicing through DNA and causing a range of damage. Exposure can increase the risk of various cancers and even cause genetic mutations that can be passed on to future generations.

Recent Breakthroughs in EMR Research

Scientists are constantly working to understand and mitigate the harmful effects of EMR. Here are some exciting developments:

Computational Modeling: Imagine being able to predict how EMR interacts with your body using sophisticated computer models. This is helping design safer medical devices and occupational safety guidelines.

Thermal Imaging: Like a heat-sensing camera, thermal imaging can detect temperature changes in tissues during EMR exposure, preventing severe burns.

Bio dosimetry: This technique allows scientists to measure the dose of ionizing radiation a person has received, helping assess individual risk and guide medical treatment.

Conclusion: Navigating the Electromagnetic Landscape

EMR is an integral part of our modern world, but it's crucial to be aware of its potential dangers. By understanding the different forms of EMR and their effects, we can take steps to protect ourselves and future generations. As technology advances, we can look forward to even more innovative solutions to mitigate the risks and harness the benefits of this powerful force.

Sound

The Symphony of Vibrations: A Sonic Adventure
Imagine a world without sound. No laughter, no music, no whispers of wind rustling through trees. Sound is the vibrant pulse of our world, a magical tapestry woven from the threads of vibration. Let's embark on a sonic adventure to uncover the secrets of how these vibrations create the symphony of life around us.

1. The Dance of Disturbance: Where Sound Begins

Every sound, from the gentle hum of a bee to the roar of a rocket launch, is born from a tiny dance – the vibration. Picture a mischievous sprite shaking a tambourine, sending shivers of energy outwards. This is what happens when any object vibrates; it nudges the surrounding air, water, or even solid ground, setting off a ripple effect. These ripples are sound waves, carrying the energy of the vibration outwards like a message in a bottle.

2. The Medium's Magic: A Sonic Highway

Think of the medium – air, water, or a solid – as a bustling highway for sound waves. When the mischievous sprite shakes its tambourine, it creates a domino effect, with air molecules bumping into each other and passing on the vibration. The density of the "traffic" on this highway affects the speed of the sound. Imagine sound racing through a tightly packed crowd (a solid) versus a leisurely stroll through a park (air).

3. The Vibrating Orchestra: A World of Sounds

Our world is a grand orchestra of vibrating sources. Nature provides the percussion section, with rustling leaves, crashing waves, and the rumble of thunder. Humans contribute the melodic instruments, from the delicate strings of a violin to the powerful brass of a trumpet. Even our own voices are instruments, with vocal cords acting as strings, controlled by the breath of our lungs.

4. The Language of Sound: Deciphering the Code

Sound waves speak a language of their own, with amplitude, frequency, and timbre as its words. Amplitude controls the volume, like a dial turning up the intensity of the vibration. Frequency determines the pitch, whether it's a high-pitched squeak or a low-pitched rumble. Timbre is the unique character of a sound, like the difference between a flute and a clarinet playing the same note.

5. Sound's Superpowers: Beyond Our Ears

Sound isn't just something we hear; it's a force with hidden powers. Ultrasound, with its super-fast vibrations, allows us to see inside the human body and even levitate tiny objects in mid-air. Noise-canceling technology uses sound waves to fight fire with fire, silencing unwanted noise. And in the realm of virtual reality, sound creates immersive experiences that transport us to other worlds.

6. Sonic Wonders: Unveiling the Mysteries

The world of sound is full of mysteries waiting to be unlocked. Why do Stradivarius violins have such a unique and beautiful sound? How do lions produce their earth-shaking roars? By exploring these questions, we delve deeper into the magic of sound and its power to connect us to the world around us.

7. The Sonic Tapestry: A Never-Ending Story

The journey into the world of sound is an ongoing adventure. As we continue to explore the intricate dance of vibrations, we uncover new secrets and applications of this fundamental force. Sound is the soundtrack of our lives, a symphony of vibrations that connects us to the past, present, and future.

The Symphony of Life: A Sonic Adventure 🎶

Imagine a world without sound. No laughter, no music, no whispers of wind rustling through leaves. A silent film with no soundtrack. Pretty bleak, right? Thankfully, our world is a vibrant orchestra of sounds, a symphony that plays out every second of our lives. Let's dive headfirst into this incredible sonic landscape!

1. Ears: Those Amazing Sound Catchers!

Our ears are like these super-sensitive antennas, picking up vibrations in the air and transforming them into something we understand – sound! But here's the catch: we can only hear a certain range of these vibrations, kind of like tuning into a specific radio station. We call this the "audible range," and it goes from super low rumbles (like a distant earthquake) to those ridiculously high-pitched squeaks that only your dog seems to notice.

Think of it like this: imagine a piano keyboard. The lowest keys are like the deep rumble of a truck engine, and the highest keys are that annoying mosquito buzz. Most of us can hear across that whole keyboard, but as we get older, it's like losing the ability to hear those really high notes. Kind of like the piano tuner forgot to come for a while!

Beyond Human Hearing: The Secret Soundscape

But guess what? The piano of sound has way more keys than we can even imagine! There are sounds so low we can't hear them – we call them "infrasound." Elephants use these to chat with each other across miles, like a super-secret elephant phone call! And then there are sounds so high they're off the charts – that's "ultrasound." Bats use this to "see" in the dark, like a built-in radar system. Pretty cool, huh?

2. Sound Needs a Ride: The Medium is the Message (Literally!)

Ever try shouting in outer space? (Okay, maybe not literally). But if you did, no one would hear you! That's because sound needs something to travel through, like air, water, or even a solid wall. Think of it like those little waves you make when you drop a pebble in a pond. The water is the "medium" that carries the wave. Sound is similar – it needs those tiny air particles to bump into each other and carry the vibration along.

The Bell Jar Experiment: Silence is Golden (and Kinda Spooky)

Want to see this in action? Grab a bell jar, a vacuum pump, and a bell (the ringing kind, not the taco kind). Put the ringing bell in the jar and suck out all the air. As the air disappears, the sound gets quieter and quieter until... silence! It's like magic, but it's really just science! This proves that without those air particles, sound has no way to travel.

3. The Sound Recipe: Amplitude, Frequency, and Our Brain's Interpretation

So, we know sound needs a medium to travel, but what makes one sound different from another? That's where "amplitude" and "frequency" come in. They're like the secret ingredients in the sound recipe.

Amplitude: The Volume Knob

Amplitude is like the volume control on your stereo. The bigger the amplitude, the louder the sound. Think of a whisper compared to a rock concert. The whisper has small sound waves, while the concert has giant ones that practically knock you over!

Frequency: The Pitch Perfect

Frequency is all about how high or low a sound is. A high-pitched squeal has lots of sound waves packed together, while a deep rumble has long, stretched-out waves. It's like the difference between a tiny flute and a giant tuba.

Our Brains: The Sound Mixer

But here's where it gets really interesting. Our brains take those amplitudes and frequencies and turn them into what we experience as "loudness" and "pitch." It's like our brains have their own built-in sound mixer! And sometimes, our brains can even play tricks on us, making us hear things differently depending on how loud or quiet a sound is.

The Mosquito Ringtone: Hear It If You Can!

Ever heard of the "Mosquito Ringtone"? It's this super high-pitched sound that only young people can usually hear. Older folks often can't pick it up because their "high-note hearing" isn't as good. It's like their inner piano is missing a few keys at the top!

Sound: It's a Wrap!

So, there you have it – a whirlwind tour of the amazing world of sound! From the way it travels to how we hear it; sound is a truly fascinating phenomenon. So next time you hear a bird sing, a dog bark, or your favorite song, take a moment to appreciate the incredible science behind it all. It's a symphony of vibrations, and we're lucky to be a part of it!

1. Sound Sources and Timbre: A Deep Dive

Imagine sound as a vibrant tapestry woven from countless threads of vibration. Each thread, a tiny pulsation of energy, dances through the air, carrying with it the essence of its source. The unique character of a sound, its timbre, is the intricate pattern formed by these dancing threads.

Think of an oscilloscope as a magical loom that captures these vibrant threads and displays them as a mesmerizing dance of waves. The intricate shapes of these waves reveal the hidden secrets of sound, allowing us to peer into the very heart of its creation.

Harmonics: The Symphony of Overtones

While a pure tone is a single, clear note, most sounds are a rich chorus of frequencies, a symphony of overtones. These overtones, or harmonics, are like the supporting voices in a choir, adding depth and richness to the melody.

The fundamental frequency is the lead singer, the voice that determines the pitch we perceive. The harmonics are the harmony singers, their voices blending with the lead singer to create a unique and colorful sound.

Oscilloscope Visualization: A Window into Sound

An oscilloscope is like a window into the soul of sound, revealing its intricate and ever-changing form. It paints a picture of sound as a dynamic dance of waves, each with its own unique shape and character.

A pure tone, like the pure note of a tuning fork, appears as a graceful sine wave, a simple and elegant curve. Complex sounds, like the rich tones of a musical instrument or the human voice, are a whirlwind of waves, a symphony of shapes and patterns.

Examples of Sound Sources and Their Timbre

The violin, with its warm and expressive voice, is a master of harmonics. Its body, like a resonating chamber, amplifies certain harmonics, creating a rich and complex tapestry of sound.

The flute, with its clear and mellow tone, is a minimalist of harmonics. Its sound is like a gentle breeze, a smooth and flowing wave with few embellishments.

The human voice, the most versatile of instruments, is capable of producing a vast range of timbres. The shape of our vocal tract and the tension of our vocal cords act as a sculptor, shaping the harmonics to create the myriad sounds of speech and song.

Factors Affecting Timbre

Timbre is a delicate balance of several factors, each contributing to the unique character of a sound.

The harmonic content is the palette of colors, the range of overtones that blend to create the overall hue.

The envelope is the shape of the sound, the way it rises and falls, swells and fades.

Vibrato is a subtle vibrato, a gentle shimmer that adds warmth and expressiveness.

Tremolo is a rhythmic pulse, a gentle throb that adds a sense of movement and life.

Case Study: Distinguishing Instruments

Our brains are like master detectives, able to identify instruments based on their unique timbral fingerprints. We can distinguish the warm, woody tone of a cello from the bright, brassy sound of a trumpet, even in the midst of a complex orchestral arrangement.

Conclusion

Timbre is the soul of sound, the unique and expressive character that sets each sound apart. It is a delicate interplay of harmonics, envelope, vibrato, and other factors, all woven together to create a rich and vibrant tapestry of sound.

2. Echoes: The Reflection of Sound

Imagine standing at the edge of a vast canyon, your voice echoing off the towering walls. The echo is a phantom of your own voice, a delayed repetition that seems to come from the depths of the canyon itself.

Echoes are like sound mirrors, reflecting sound waves back to their source. They are a reminder that sound is not just a fleeting vibration, but a tangible entity that can interact with its environment.

Mechanism of Echo Formation

When a sound wave encounters a surface, it's like a ball bouncing off a wall. Some of the sound energy is reflected back, while the rest is absorbed or transmitted through the material.

The reflected wave travels back towards the source, and if the delay is long enough, we perceive it as a distinct echo. It's like hearing your own voice calling back to you from a distance.

Factors Affecting Echo Perception

Distance is key to echo perception. The farther the sound source is from the reflecting surface, the longer the delay and the more noticeable the echo.

Surface properties also play a role. Hard, smooth surfaces like cliffs or walls are like perfect mirrors, reflecting sound with high efficiency. Soft, porous materials like carpets or curtains are like sound absorbers, muffling the reflection.

Sound intensity matters too. Louder sounds have more energy to be reflected, making their echoes more prominent.

The listener's position is also important. Standing closer to the reflecting surface or farther from the sound source can enhance echo perception.

Applications of Echoes

Echoes are not just a natural phenomenon; they have found numerous applications in various fields.

Sonar systems use echoes to explore the underwater world, like bats using echolocation to navigate in the dark.

Architects use echoes to design concert halls and auditoriums, creating spaces where sound can reverberate and enhance musical performances.

Animals like bats and dolphins use echolocation to navigate and hunt, their clicks and whistles bouncing off objects to reveal their location.

Musicians use echoes creatively, adding depth and spaciousness to their compositions.

Case Study: The Grand Canyon

The Grand Canyon, with its vast expanse and towering walls, is a natural amphitheater for echoes. A single shout can trigger a cascade of echoes, bouncing back and forth between the canyon walls.

Conclusion

Echoes are a fascinating reminder that sound is a tangible entity that can interact with its environment. They are used in various fields, from navigation and architecture to music and animal behavior, showcasing the versatility and importance of sound reflection.

3. Simple Experiments to Demonstrate Sound Reflection

Sound reflection is not just a theoretical concept; it can be easily demonstrated through simple experiments using everyday materials. These experiments bring the magic of sound to life, allowing us to experience and visualize the phenomenon of sound reflection.

Experiment 1: Two Tubes and a Wall

Imagine two cardboard tubes, like secret tunnels for sound, placed end-to-end against a wall. A sound source at one end of the tubes sends sound waves on a journey through the tubes, bouncing off the wall and emerging at the other end. It's like a game of sound hide-and-seek, with the wall acting as a hidden mirror.

Experiment 2: Mirror and a Watch

A concave mirror, like a sound magnifier, can focus sound waves, amplifying the ticking of a watch placed in front of it. It's like listening to the heartbeat of time, amplified by the curved surface of the mirror.

Experiment 3: Cardboard and a Protractor

A piece of cardboard becomes a sound deflector, bouncing sound waves at different angles. A protractor measures the angles, revealing the precise relationship between the angle of incidence and the angle of reflection. It's like a game of sound billiards, with the cardboard acting as a bumper and the sound waves as the balls.

Experiment 4: Echo in an Empty Room

An empty room becomes an echo chamber, amplifying the sound of a clap or shout. The hard surfaces of the walls act like acoustic mirrors, reflecting sound waves back and forth, creating a chorus of echoes.

Conclusion

These simple experiments bring the magic of sound reflection to life, allowing us to experience and visualize the phenomenon firsthand. They demonstrate that sound is not just a fleeting vibration, but a tangible entity that can interact with its environment in fascinating ways.

Sound's Speedy Secrets: A Sonic Adventure

Imagine sound as a mischievous sprite, zipping through the air at different speeds depending on its mood – or rather, the temperature! It's not just a fixed number like we often learn; it's a value that changes like the weather. Let's dive into the fascinating world of sound speed and discover how to measure its elusive pace.

Chasing Sound: A DIY Experiment

Ever tried catching sound? It's like trying to grab a handful of smoke! But we can cleverly measure its speed with a simple experiment. Picture this:

The Setup: A vast open field, like a giant playground for sound. You'll need a loud noisemaker (think clapping hands, a starting pistol, or even two wooden blocks clacking together), a super accurate stopwatch, and a long measuring tape.
The Chase: You and a friend stand far apart, like two trees in the field. One of you makes the sound – BOOM! – and the other starts the stopwatch the very instant they see the action. Then, they stop it the moment they hear the sound.

The Calculation: Divide the distance between you by the time on the stopwatch. Voilà! You've just calculated the speed of sound!

Sound's Quirks: Why It's Never the Same

Sound's speed isn't just a random number; it's affected by a few key things:

Temperature Tango: Heat makes air molecules dance faster, and sound loves to join the party! The hotter it is, the quicker sound travels. It's like sound gets a boost of energy from the warmth.

Humidity's Helping Hand: Ever notice how sound seems to travel further on a foggy day? That's because humid air is less dense, giving sound a slightly easier path to follow.

Air's Unique Blend: Even the mix of gases in the air can subtly change sound's speed. It's like sound has a preference for certain air flavors!

Sound's Superpowers in Different Worlds

Sound isn't just limited to the air; it can travel through anything that vibrates. But it's a bit of a shape-shifter, changing its speed depending on the environment:

Speedy Solids: Sound loves to zoom through solids, like a superhero with super speed! That's because the molecules in solids are tightly packed, allowing sound to jump quickly from one to the next. Think of a whisper traveling down a long metal pipe.

Liquid Laziness: Sound slows down in liquids, like it's swimming through water. The molecules are more spread out, making it a bit harder for sound to move. Imagine listening to music underwater – it sounds different, right?

Gaseous Go-Slow: Sound takes its time in gases, like it's strolling through a park. The molecules are far apart, so sound has to work harder to travel. That's why we hear things almost instantly in a room, but it takes a few seconds for thunder to reach us after a lightning strike.

Sound's Amazing Feats

Sound's ability to travel at different speeds has some incredible applications:

Underwater Adventures: Sonar systems use sound to "see" underwater, like dolphins and whales! By bouncing sound waves off objects, we can map the ocean floor and find hidden treasures.
Earth's Secrets: Earthquakes create sound waves that travel through the Earth, revealing its hidden layers like an X-ray.
Body Scans: Ultrasound uses sound waves to create pictures of our insides, helping doctors diagnose and treat illnesses.
Musical Magic: The speed of sound is what makes musical instruments work! Different materials and lengths create different sounds, from the deep rumble of a cello to the high-pitched trill of a flute.
Sound-Scaping: Architects use their knowledge of sound speed to design buildings that sound amazing, whether it's a concert hall with perfect acoustics or a library where whispers fade away.
The Sonic Symphony

Sound is more than just something we hear; it's a dynamic force that shapes our world. By understanding its speed and how it changes, we can unlock its secrets and harness its power for amazing things. So next time you hear a sound, take a moment to appreciate the incredible journey it's taken to reach your ears!

Ultrasound: Seeing the Unseen with Sound

Imagine a world where we could peer inside objects and bodies without ever making a cut. That's the magic of ultrasound, a technology that uses sound waves to reveal hidden wonders, from the delicate gears of a watch to the flutter of a tiny heart.

Ultrasound Cleaning: A Symphony of Bubbles

Think of ultrasound cleaning like a microscopic bubble bath for your valuables. High-frequency sound waves create tiny bubbles in a liquid, and when these bubbles burst, they gently lift away dirt and grime from even the most intricate surfaces.

Jewelry: Ultrasound can make your diamonds sparkle like never before, reaching into tiny crevices that brushes can't touch.

Medical Instruments: Ultrasound ensures surgical tools are pristine, using sound instead of harsh chemicals to eliminate any lingering germs.
Electronics: Say goodbye to dust bunnies hiding in your gadgets! Ultrasound can clean delicate circuitry without causing any damage.

Medical Scanning: A Window into the Body

Ultrasound has revolutionized healthcare, allowing doctors to see inside the body with incredible clarity and safety.

Prenatal Peek: Expectant parents can marvel at their baby's tiny features and movements during an ultrasound scan, a truly magical experience.
Heart Health: Echocardiograms use ultrasound to create moving pictures of the heart, helping doctors diagnose and monitor heart conditions.
Body Exploration: Ultrasound can visualize organs like the liver, kidneys, and thyroid, aiding in the diagnosis of various conditions.

Sonar: Exploring the Depths with Sound

Sonar, short for Sound Navigation and Ranging, is like an underwater echolocation system. By emitting sound waves and listening for their echoes, we can map the ocean floor, locate shipwrecks, and even track schools of fish.

Advanced Ultrasound: Pushing the Boundaries

Ultrasound technology continues to evolve, offering even more amazing possibilities:

3D and 4D Ultrasound: Experience the wonder of seeing your unborn baby in three dimensions, complete with moving images in 4D!
Doppler Ultrasound: This technique allows doctors to visualize blood flow, helping diagnose circulatory problems.
HIFU: High-Intensity Focused Ultrasound uses sound waves to target and destroy tumors, a non-invasive alternative to surgery.

Ultrasound: A Sound Investment in the Future

From cleaning and medical imaging to underwater exploration, ultrasound is a versatile and indispensable tool. As technology advances, we can expect even more incredible applications of this remarkable technology, shaping our world in ways we can only imagine.

Electricity and magnetism

Simple magnetism and magnetic fields

The Magic of Magnets: A Whimsical Journey into the World of Invisible Forces

Imagine a world where things could stick together without glue, trains could fly, and invisible shields could protect us from harm. Sounds like something out of a fantasy novel, right? Well, this is the world of magnets, where the impossible seems possible, and the invisible forces at play are nothing short of magical.

The Dance of Attraction and Repulsion

At the heart of this magic lies a simple dance: the dance of attraction and repulsion. Like two shy dancers, the north and south poles of magnets can't resist each other's pull. But bring two north poles or two south poles together, and it's like trying to force those shy dancers into a head-on collision – they'll push each other away with all their might.

This push and pull, known as magnetic force, is what makes magnets so fascinating. It's the invisible hand that guides a compass needle to point north, allows a refrigerator magnet to cling to its metal door, and enables powerful magnets to lift heavy objects with ease.

The Invisible Aura: Magnetic Fields

Every magnet is surrounded by an invisible aura, a sort of magnetic force field that extends outwards like an invisible embrace. This aura, called a magnetic field, is what allows magnets to interact with each other and with other magnetic materials, even without touching.

Think of it like a dancer's swirling dress. The closer you get to the dancer, the stronger the swirl of the dress. Similarly, the closer you get to a magnet, the stronger its magnetic field. And just like the dress can influence objects around it, the magnetic field can exert forces on other magnets and magnetic materials.

The Art of Induced Magnetism

But the magic of magnets doesn't stop there. Magnets have the power to temporarily bestow their magnetic properties onto other materials, like a fairy godmother granting a magical gift. This phenomenon, known as induced magnetism, is what allows a magnet to pick up a chain of paperclips, each clip becoming a temporary magnet itself, clinging to the one before it.

Magnets in Action: From Compasses to MRI Machines

The magic of magnets is not just a source of wonder and amusement; it's also a powerful force that has shaped our world in countless ways. From the humble compass that guided explorers across vast oceans to the powerful MRI machines that allow doctors to peer inside the human body, magnets have played a crucial role in our technological advancements.

The Future of Magnets: A World of Endless Possibilities

And the journey of magnets is far from over. Scientists are constantly exploring new ways to harness the magic of magnets, from developing new magnetic materials with extraordinary properties to creating innovative technologies like magnetic levitation trains that glide effortlessly above the tracks.

So, the next time you encounter a magnet, take a moment to appreciate the magic it holds. It's a world of invisible forces, where attraction and repulsion dance in perfect harmony, and where the impossible becomes possible. It's a world that continues to inspire and amaze, reminding us that the most magical things are often hidden in plain sight.

The Curious Case of the Clingy Metal: A Magnetic Mystery!

Ever played with a magnet and wondered why it sticks to some things but completely ignores others? It's like the magnet has a secret club, and only certain materials get an invite! Let's unravel this mystery and dive deep into the world of magnetic and non-magnetic materials.

Imagine tiny dancers inside every material. These dancers are actually electrons, and they're constantly spinning, creating mini magnetic fields. In most materials, these dancers pair up, spinning in opposite directions – like a perfectly choreographed waltz where all the spins cancel each other out. No net magnetic field, no party invite!

But in magnetic materials, some of these dancers like to go solo! Their unpaired spins create a tiny magnetic field, and when these tiny fields join forces, the material becomes a magnet. Think of it like a flash mob of electron dancers, all spinning in the same direction and creating a magnetic frenzy!

Think of a magnet like a popular kid in school. Everyone wants to be their friend! Magnetic materials are like the eager students who rush towards the popular kid, while non-magnetic materials are content to hang back and do their own thing.

Here's a quick rundown of who's in and who's out of the magnet's exclusive club:

Magnetic Materials (The "In" Crowd) Non-Magnetic Materials (The Independent Crew)
Response to a Magnet Can't resist a magnet's pull! "Meh, magnets are overrated."
Electron Dancers Flash mob! All spinning in the same direction. Perfectly paired waltzes, canceling each other out.
Examples Iron (the ultimate magnet lover!), nickel, cobalt, steel Wood, plastic, rubber, copper, gold, silver

Export to Sheets
But wait, there's more! Just like there are different cliques in school, there are different types of magnets:

Temporary magnets are like that friend who's super enthusiastic at first but quickly loses interest. They lose their magnetism easily. Think of soft iron – great for electromagnets, but not for sticking your artwork to the fridge.

Permanent magnets are the loyal best friends, sticking with you through thick and thin (and even when you take away the external magnetic field!). Steel, alnico, and neodymium-iron-boron are the superstars here, used in everything from loudspeakers to motors.

Here's a fun fact: Did you know that the Earth itself is a giant magnet? That's why compasses work! And speaking of compasses, ancient sailors used lodestone, a naturally occurring magnetic material, to navigate the seas.

From the simple compass to the powerful MRI machine, magnetism plays a vital role in our lives. It's a force that both attracts and fascinates, and its mysteries continue to inspire scientists and inventors to explore its potential. So, the next time you play with a magnet, remember the tiny dancers within, and marvel at the invisible forces that shape our world!

Imagine a world where invisible forces dance and twirl, shaping the reality around us. This is the world of magnetic fields, a realm of unseen power that affects everything from the humble compass to the intricate workings of our planet.

Think of a magnet as a tiny superstar, radiating an aura of influence. This aura, the magnetic field, is a zone where other magnets or moving electric charges feel a push or a pull. It's like an invisible handshake, sometimes friendly, sometimes a bit repellant, depending on the magnetic personalities involved.

To visualize this hidden world, we use magnetic field lines, like trails of glitter tracing the steps of an invisible dancer. These lines reveal the direction and strength of the magnetic field. Where the lines cluster close, the field flexes its muscles; where they spread out, it takes a breather.

These lines are also quite the rule-followers:

Always on the move: They form continuous loops, flowing from the north pole of a magnet to its south, like a never-ending conga line.
Never crossing paths: Imagine the chaos if these lines bumped into each other! It would be like a compass trying to point in two directions at once.

We can even map these invisible lines, like explorers charting a new territory:

The Compass, our trusty guide: A tiny compass needle, like a miniature detective, always points in the direction of the magnetic field line. By following its lead, we can trace the lines and map the field.
Iron Filings, the eager followers: Sprinkle these tiny particles over a magnet, and they'll snap into formation, revealing the magnetic field lines like a hidden pattern suddenly brought to life.
Magnetic fields are everywhere, shaping our world in countless ways:

Earth's magnetic embrace: Our planet itself is a giant magnet, its field acting like a protective shield against harmful solar radiation.
The dance of electricity and magnetism: In electric motors and generators, magnetic fields and moving charges engage in a graceful tango, converting electrical energy into motion and vice versa.
Medical marvels: MRI machines use powerful magnetic fields to peer inside the human body, creating detailed images that help doctors diagnose and treat illnesses.
Floating trains: Maglev trains, propelled by magnetic forces, glide effortlessly above the tracks, offering a glimpse into a future of high-speed transportation.
The magnetic field, a silent force, a hidden choreographer, shaping the world around us in ways we are only beginning to understand. It's a testament to the beauty and complexity of the universe, a reminder that there's always more to discover, more to explore, in the vast expanse of the unseen.

Journey into the Realm of Magnets: Where the Invisible Holds Power

Imagine a world where the invisible dances and pulls, where forces we can't see shape our everyday lives. This is the realm of magnets, a world of mystery and wonder that has captivated scientists and dreamers for centuries. Let's embark on a journey to explore this fascinating domain, where we'll unravel the secrets of magnetic fields, delve into the magic of permanent magnets, and harness the power of electromagnets.

1. The Magnetic Compass: Nature's Tiny Guide

Picture a tiny compass needle, delicately balanced and always pointing north. This simple device, used for centuries by explorers and adventurers, holds the key to understanding the direction of a magnetic field. Imagine the magnetic field as a series of invisible lines flowing from the north pole of a magnet to its south pole. When you place a compass near a magnet, its north-seeking pole aligns itself with these lines, revealing the direction of the magnetic force.

2. Magnetic Field Lines: A Dance of Attraction and Repulsion

Imagine iron filings sprinkled around a magnet, magically arranging themselves into intricate patterns. These patterns, formed by magnetic field lines, provide a visual representation of the magnetic field's strength and direction. The closer the lines, the stronger the magnetic force, like a bustling crowd drawn together by an irresistible attraction. Conversely, as the lines spread apart, the force weakens, like a fading whisper in the distance.

3. Permanent Magnets: Nature's Gift of Magnetism

From the humble refrigerator magnet holding your child's artwork to the powerful magnets used in medical imaging, permanent magnets are ubiquitous in our lives. These remarkable materials possess a unique ability to retain their magnetism over long periods, providing a constant source of magnetic force without the need for external power.

Real-World Wonders of Permanent Magnets:

Data Storage: Imagine tiny magnets embedded in your computer's hard drive, storing your precious photos, videos, and documents as intricate magnetic patterns.
Navigation: Picture a sailor navigating the vast ocean using a compass, its magnetized needle guided by the Earth's magnetic field, a beacon in the boundless expanse.
Medical Marvels: Envision powerful magnets in MRI machines, peering into the human body, revealing its intricate details, and aiding in the diagnosis and treatment of diseases.

Sound and Music: Imagine magnets in your headphones, transforming electrical signals into the melodies and rhythms that move your soul.

Case Study: Magnets in Wind Turbines

Imagine towering wind turbines harnessing the power of the wind, their blades gracefully turning to generate clean energy. At the heart of these turbines lie powerful permanent magnets, working tirelessly to convert wind energy into electricity. These magnets have revolutionized wind power, making it a more efficient and sustainable energy source.

4. Electromagnets: Taming the Magnetic Force

Imagine a simple coil of wire wrapped around an iron core, transformed into a powerful magnet with the flick of a switch. This is the magic of electromagnets, magnets that can be turned on and off, their strength controlled by the flow of electric current.

Electromagnets in Action:

Industrial Giants: Picture massive electromagnets in scrapyards, effortlessly lifting and moving heavy objects, like metal giants playing with their toys.
High-Speed Travel: Envision Maglev trains gliding silently above the tracks, propelled by powerful electromagnets, a futuristic vision of transportation.
Medical Miracles: Imagine electromagnets in MRI machines, generating strong magnetic fields that allow doctors to see inside the human body with unprecedented clarity.
Everyday Essentials: Think of the humble doorbell, its chime activated by an electromagnet, a simple yet essential part of our daily lives.

Case Study: Electromagnets in Magnetic Resonance Imaging (MRI)

Imagine lying inside an MRI machine, its powerful electromagnets generating a magnetic field that interacts with your body's atoms, creating detailed images of your organs and tissues. This remarkable technology has revolutionized medical diagnostics, allowing doctors to detect and treat a wide range of diseases.

In Conclusion

From the tiniest compass to the most powerful MRI machine, magnets shape our world in countless ways. They are the invisible force that guides our navigation, stores our data, powers our industries, and heals our bodies. As we continue to explore the mysteries of magnetism, we can expect even more incredible discoveries and applications in the future.

Electrical quantities

Electrical charge

The Quirky Universe of Electric Charge: A Tale of Two Forces
Imagine the universe as a grand cosmic dance floor, where everything is grooving to the rhythm of four fundamental forces. There's gravity, the wallflower that keeps everyone grounded. The strong and weak forces are the close-knit couple, always interacting in their own little world. But then there's electromagnetism – the life of the party, flashing and sparking, responsible for everything from the dazzling aurora borealis to the gentle hum of your refrigerator.

And what's the secret ingredient that makes electromagnetism so electrifying? It's electric charge, the tiny but mighty force that governs how particles interact. Think of it like personality: some particles are positive, always radiating good vibes, while others are negative, a bit more reserved.

Opposites Attract, Likes Repel: A Cosmic Dance-Off

Just like in a high school dance, opposites attract in the world of electric charge. Positive charges can't help but be drawn to negative charges, while like charges push each other away like awkward dance partners. It's a constant push and pull, a never-ending tango of attraction and repulsion.

The Coulomb: A Unit with Shocking Proportions

Now, how do we measure this electric "personality"? We use a unit called the coulomb, named after the dapper French physicist Charles-Augustin de Coulomb. One coulomb is a seriously hefty amount of charge – picture a lightning bolt zapping down from the sky, and you're only looking at about 15 coulombs! But don't worry, most of the time we're dealing with tiny charges, like the microcoulombs buzzing around in your phone battery.

The Elementary Charge: The Atom's Tiny Dance Crew

Deep down inside atoms, we meet the smallest dancers of them all: protons and electrons. Protons, the positively charged party animals, hang out in the atom's nucleus, while electrons, the negatively charged introverts, whirl around them like a shy but energetic cloud. And the amazing thing? Their charges are perfectly balanced, always an integer multiple of the elementary charge, the smallest unit of charge ever discovered. It's like the universe has a strict dress code for its electric dance party!

Coulomb's Law: The Dance Floor Rules

But how exactly do these charges interact? Enter Coulomb's Law, the rulebook for this electric dance-off. It tells us that the closer two charges get, the stronger their attraction or repulsion becomes. It's like the electric version of "personal space" – the closer you get, the more intense the interaction.

Electric Charge in Action: From Photocopiers to Lightning Bolts

This electric dance isn't just happening in some abstract realm – it's all around us! When you rub a balloon on your hair and it sticks to the wall, that's static electricity in action. When a thunderstorm cracks the sky with a blinding flash, that's a massive discharge of electric charge seeking equilibrium. Even the humble photocopier relies on the magic of electrostatics to create copies.

Beyond the Basics: The Electric Boogie Continues

The world of electric charge is a vast and vibrant one, full of surprises and mysteries. From the intricate dance of electrons in a circuit to the mind-bending concept of electromagnetic waves, there's always something new to discover. So put on your dancing shoes, embrace the electric boogie, and get ready to explore the fascinating world of electric charge!

The Spark of Life: A Whimsical Journey into Electrostatic Charging by Friction

Have you ever felt the zap of a doorknob after walking across a carpeted floor? Or witnessed the magic of a balloon sticking to your hair after a vigorous rub? These everyday wonders are all thanks to the enchanting phenomenon of electrostatic charging by friction, or as I like to call it, the "dance of the electrons."

Imagine tiny, mischievous electrons leaping from one material to another when they get close enough to tango. This playful exchange leaves one material with a positive charge, like a sunbeam, and the other with a negative charge, like a moonbeam. And just like the sun and moon, these opposite charges can't help but attract!

Let's explore this electrifying world through a few captivating experiments:

1. The Balloon's Captivating Embrace:

Picture a vibrant balloon, yearning for connection. As you rub it against your hair, it's like whispering secrets to the strands, coaxing electrons to jump onto the balloon's surface. Suddenly, the balloon becomes a magnet, playfully pulling your hair or tiny scraps of paper into its embrace. It's a delightful dance of attraction!

2. The Comb's Aquatic Ballet:

Imagine a thin stream of water gracefully flowing from a faucet. Now, introduce a comb that's been energized by running it through your hair. As the comb approaches the water, the water bends towards it, as if bowing to a mesmerizing conductor. This captivating ballet is a testament to the invisible forces at play.

3. The Electroscope's Dramatic Revelation:

This experiment unveils the secrets of charge with a touch of drama. An electroscope, with its delicate gold leaves, acts as a stage. When a charged object approaches, the leaves dramatically diverge, like startled dancers leaping apart. This captivating performance reveals the presence of invisible charges.

4. The Wimhurst Machine's Electric Symphony:

Prepare to be amazed by the Wilmshurst machine, a grand contraption that generates miniature lightning bolts! As you crank its handle, the machine whirs to life, creating a mesmerizing spectacle of sparks jumping across its metal spheres. It's a symphony of crackling energy, showcasing the power of electrostatic forces.

But why do only electrons get to play this game of tag?

Well, it's all about their position in the atomic world. Electrons, those nimble particles, orbit the atom's nucleus like planets around a star. They're less tightly bound than their nuclear counterparts, the protons and neutrons, making them free to embark on exciting adventures.

Think of the triboelectric series as a popularity contest for electrons. Materials at the top are like social butterflies, eager to shed their electrons and become positively charged. Those at the bottom are more introverted, happily accepting electrons and becoming negatively charged.

The Spark of Inspiration:

Electrostatic charging isn't just a source of amusement; it's a fundamental force shaping our world. From the awe-inspiring power of lightning to the everyday convenience of photocopiers, this phenomenon touches our lives in countless ways.

So, the next time you witness a static shock or see a balloon clinging defiantly to a wall, remember the enchanting dance of the electrons. It's a reminder that even the smallest particles can create extraordinary magic.

The Electric Field: An Invisible Dance of Attraction and Repulsion

Have you ever felt a spark, a crackle in the air before a thunderstorm? That's the electric field, an invisible forcefield that surrounds every charged object, like an aura of unseen energy. Imagine it as a cosmic dance floor where charged particles boogie down, pulling and pushing each other with an invisible force.

1. Spooky Action at a Distance

Picture a lone electron, a tiny dancer in the vastness of space. Even though it's all alone, it throws out an invisible net, its own personal dance floor, called an electric field. Now, introduce another charged particle – it's like a new dancer stepping onto the floor. Suddenly, they feel the music, the rhythm of the electric field, and they can't help but move! If they have the same charge, they'll push each other away in a fierce tango. Opposite charges? They'll be drawn together in an electrifying waltz.

Key Takeaways:

No Touching! The electric field is like telekinesis for charged particles. They can interact without ever actually touching, like expert dancers who know how to move in perfect sync.
Invisible Choreography: The electric field is a vector field, meaning it has both strength and direction. It's like a choreographer, telling each charged particle how strongly to move and in what direction.
Born to Dance: Every charged particle, whether it's moving or standing still, creates its own electric field. It's like their own unique dance style that influences everyone around them.

2. Following the Electric Slide

The direction of the electric field is like an arrow pointing the way for a positive charge. Imagine a tiny explorer with a positive charge – the electric field tells him which way to move. This "electric slide" is visualized through electric field lines:

From Plus to Minus: Field lines flow out of positive charges (like rays of sunshine) and into negative charges (like a welcoming embrace).
No Crashing Allowed: Field lines never cross, just like dancers on a crowded floor. Can you imagine the chaos if they did?
Crowd Surfing: The closer the field lines, the stronger the electric field, like a mosh pit of charged particles.

Visualizing the Electric Field:

Imagine a positively charged disco ball, radiating lines of electric force outwards, pushing away any other positive charges that come too close. A negatively charged ball, on the other hand, is like a black hole, pulling positive charges in with its irresistible charm.

Example:

Picture a positive and negative charge, like two dancers locked in a passionate embrace. The electric field lines would flow out of the positive charge, curve gracefully towards the negative charge, and finally sink into it, showing the irresistible attraction between them.

3. The Electric Field's Mathematical Moves

The electric field has its own secret language – mathematics! It's represented by the symbol 'E' and its strength is calculated using this formula:

$E = F/q$

Think of it like this:

E (Electric Field Strength): How intense the electric "music" is.
F (Force): How strongly a charged particle is pushed or pulled on the dance floor.
q (Charge): How much "dance energy" the particle has.
This equation shows that the stronger the force or the smaller the charge, the more intense the electric field.

Coulomb's Law: The Electric Tango

Coulomb's Law is like the electric tango, describing how two charged particles interact:

$E = kQ/r^2$

Here's the breakdown:

k (Coulomb's Constant): The "rhythm" of the electric interaction, a fixed number.
Q (Source Charge): The main dancer, determining how strong their influence is.
r (Distance): How close the dancers are to each other. The closer they are, the stronger the attraction or repulsion.
This equation shows that the electric field gets weaker the further away you are from the source charge.

4. Electric Field Party: Multiple Charges!

When multiple charges get together, it's an electric field party! The total electric field is like the combined "music" from all the dancers, with each charge contributing its own moves. To find the total field, we add up the individual fields from each charge, taking into account both their strength and direction.

5. The Electric Field's Greatest Hits

The electric field isn't just a theoretical concept – it's a superstar with a long list of hits in the real world:

Electronics: Electric fields are the backbone of our gadgets, making our phones, computers, and TVs come to life.
Electromagnetism: Electric and magnetic fields are like two sides of the same coin, creating the electromagnetic waves that bring us radio, TV, and Wi-Fi.
Medicine: Doctors use electric fields to peek inside our bodies with technologies like ECG and EEG, which monitor the electrical activity of our hearts and brains.
Material Science: Scientists use electric fields to manipulate materials at the atomic level, creating new technologies and materials with amazing properties.
Environmental Science: Electric fields help clean up our environment by removing pollutants from industrial emissions.

Case Studies: The Electric Field in Action

Capacitor: The Electric Piggy Bank: A capacitor is like a tiny electric piggy bank, storing energy in its electric field. It has two conductive plates separated by an insulator, and when a voltage is applied, an electric field builds up between the plates, storing energy like a coiled spring.
Electrostatic Precipitator: The Air Purifier: Electrostatic precipitators use electric fields to clean up dirty air, like a giant electric broom. They charge pollutants and then attract them to collecting plates, leaving the air fresh and clean.
Electrophoresis: The Molecular Sorter: Electrophoresis is like a molecular sorting machine, using an electric field to separate molecules based on their charge and size. It's like a race where different molecules move at different speeds, allowing scientists to identify and study them.

Conclusion: The Electric Field's Encore

The electric field is a fundamental force of nature, a hidden choreographer that orchestrates the interactions of charged particles. From the tiniest atom to the largest galaxy, the electric field plays a vital role in shaping our universe. As we continue to explore its mysteries, we can expect even more exciting discoveries and innovations in the future.

Electric Fields: A Dance of Invisible Forces

Imagine a world where you could see the invisible push and pull between charged objects. This hidden world is the realm of electric fields, a symphony of forces that governs everything from the spark of static electricity to the lightning bolts that crackle across the sky.

Electric Field Lines: Mapping the Invisible

Think of electric field lines as pathways for these invisible forces. They flow outward from positive charges like rays of sunshine, and inward towards negative charges like rivers flowing towards the sea. The closer the lines, the stronger the force, like a crowded dance floor where the energy is palpable.

Conductors: The Smooth Movers

Conductors are the social butterflies of the electrical world, allowing charges to flow freely through them like a bustling crowd moving effortlessly through a wide-open doorway. Metals like copper and silver are the life of the party, their electrons readily hopping from atom to atom, creating a current that powers our homes and devices.

Insulators: The Wallflowers

Insulators, on the other hand, are the introverts, holding their electrons tightly bound to their atoms like shy guests clinging to the edges of the dance floor. Rubber, glass, and plastics are masters of isolation, preventing the flow of charge and keeping electricity safely contained.

The Electric Field's Grand Performance

Let's take a closer look at the electric field's captivating performance in different settings:

Around a Point Charge: Imagine a tiny, charged dancer spinning in the middle of the floor, their energy radiating outwards in all directions. This is the electric field around a point charge, a simple yet elegant dance of attraction and repulsion.

Around a Charged Conducting Sphere: Now picture a group of dancers huddled together, their charges evenly distributed like a perfectly synchronized ensemble. This is the electric field around a charged conducting sphere, a picture of balance and harmony.

Between Two Oppositely Charged Parallel Conducting Plates: Imagine two lines of dancers facing each other, their movements creating a mesmerizing pattern of attraction and repulsion. This is the electric field between two oppositely charged parallel conducting plates, a captivating dance of opposites.

The Ever-Evolving World of Electric Fields

Scientists are constantly exploring the depths of this invisible world, discovering new materials and phenomena that challenge our understanding of electric fields. From topological insulators with their unique conducting surfaces to high-k dielectrics that enable smaller and more powerful electronics, the future of electric fields is full of exciting possibilities.

A World of Wonder

So, the next time you encounter a spark of static electricity or witness the power of a lightning storm, remember the intricate dance of electric fields, a hidden world of forces that shapes our universe in countless ways.

The Great Electron Escape: An Electrifying Adventure!

Imagine a bustling city filled with tiny citizens called electrons. In some parts of the city, these electrons are free spirits, zooming around as they please. These are the conductors, like the bustling copper wires that bring electricity to your home. They're the life of the party, always ready to pass along the energy!

But in other areas, the electrons are stuck at home, unable to roam freely. These are the insulators, like the rubber coating on those same wires, keeping the electricity contained and you safe from harm. They're the quiet neighbors, making sure things stay calm and orderly.

Let's build an "Electron Raceway" to see this in action!

What you'll need:

A battery (think of it as the electron pump!)
A battery holder (to keep our pump steady)
A light bulb (our finish line!)
A bulb holder (to keep our finish line shining bright)
Connecting wires with alligator clips (the racetrack!)
A variety of materials to test:
Copper wire (a superhighway for electrons!)
Aluminum foil (another speedy track!)
Plastic ruler (a roadblock for electrons!)
Wooden stick (another electron obstacle course!)
Rubber band (a bouncy barrier!)

Glass rod (a slippery slope for electrons!)
Ceramic tile (a solid wall!)

Race Day!

Build your racetrack: Connect the battery, bulb, and wires, leaving a gap in the track.
Test the materials: Place each material in the gap, one by one. Does the bulb light up? If so, our electrons have made it to the finish line! That material is a conductor! If not, the electrons are stuck – it's an insulator!

Why the difference?

It all comes down to how tightly the electrons are held by the atoms in each material. In conductors, the electrons are like free-range chickens, roaming wherever they please. In insulators, they're more like caged birds, unable to escape their atomic homes.

Think of it like this:

Copper: A bustling city with free-flowing traffic. Electrons zoom around like tiny race cars!
Rubber: A quiet neighborhood with fences around every house. Electrons are stuck at home!
Real-world applications:

Electrical wiring: Copper's free-flowing electrons make it perfect for carrying electricity throughout our homes.
Insulating materials: Rubber, plastic, and glass keep us safe by preventing electricity from flowing where it shouldn't.

Beyond the basics:

Temperature: Just like hot weather can make roads melt and traffic slow down, high temperatures can affect a material's conductivity.
Impurities: Imagine potholes on our electron racetrack! Impurities in a material can hinder the flow of electrons.
Superconductors: These are like magical highways where electrons can travel at lightning speed with no resistance! They're still being

explored, but they have the potential to revolutionize technology as we know it.
The world of electricity is full of exciting discoveries! So, grab your materials, put on your safety goggles, and get ready to explore the amazing race of electrons!

Electrical current

Imagine a bustling city at rush hour. Cars, buses, and bikes weave through the streets, each carrying its passengers to their destinations. This constant flow of movement is like the electric current, a stream of tiny charged particles called electrons rushing through a wire.

Think of these electrons as commuters, eager to reach their workplace – the positive terminal of a battery or power source. The force that drives them forward is the electric field, like the urge to get to work on time. But their journey isn't smooth. They bump into obstacles along the way, like metal ions that stand in their path like slow-moving pedestrians. These collisions slow them down, but the electric field keeps pushing them forward, resulting in a steady drift of electrons.

This "electron rush hour" is what we call electrical conduction in metals. The wires in your home are like highways for these tiny commuters, delivering electrical energy to your lights, appliances, and devices. The better the conductor, the smoother the flow of traffic, just like a well-maintained highway allows for faster travel.

But what makes some materials better conductors than others? Think of it like this:

Temperature is like road construction. As the temperature rises, the metal ions vibrate more vigorously, like roadblocks that disrupt the flow of traffic. This makes it harder for the electrons to move, reducing the conductivity.
Impurities are like potholes. They disrupt the smooth flow of electrons, scattering them like a car hitting a pothole.
Different materials have different "traffic rules." Some metals, like silver and copper, have wide "lanes" and clear "signage," allowing for a smooth and efficient flow of electrons. Others, like aluminum, have

narrower "lanes" and more "obstacles," making the journey a bit more challenging.
This analogy helps us understand why metals are such good conductors. Their unique atomic structure provides a "sea" of free electrons, ready to move at a moment's notice. This makes them ideal for carrying electrical energy, powering our world in countless ways.

So next time you flip a switch or charge your phone, remember the bustling city of electrons at work, diligently carrying the energy that makes our modern lives possible.

The Current That Flows Through Our Lives: DC vs. AC

Imagine electricity as a bustling city. In this city, electrons are the busy commuters rushing to and for. Direct current (DC) is like a steady stream of cars all heading in the same direction down a one-way street. It's predictable, reliable, and gets the job done. Think of your phone battery – that's DC power keeping your digital life humming.

Alternating current (AC), on the other hand, is like a vibrant two-way street with traffic flowing in both directions, switching rhythmically. This back-and-forth flow is the key to AC's ability to travel long distances efficiently, bringing power to our homes and industries.

DC: The Steady Worker Bee

One-Way Flow: Electrons in DC always travel in the same direction, like a loyal worker bee heading straight to the hive.
Constant Polarity: Just like a compass always points north, DC has a fixed positive and negative side.
Powering Our Gadgets: DC is the lifeblood of most electronics. Your laptop, your TV remote, even your electric toothbrush – they all rely on DC's steady flow.

AC: The Long-Distance Traveler

Two-Way Flow: Electrons in AC alternate their direction, like dancers gracefully swaying to a rhythm.

Changing Polarity: The positive and negative poles in AC switch places periodically, creating a wave-like pattern.
Electrifying Our World: AC is the champion of long-distance power transmission. It's what allows electricity to travel from power plants to our homes, powering our lights, appliances, and everything in between.

A Tale of Two Currents: Finding the Right Fit

DC and AC are like two specialized tools in an electrician's toolbox. Each has its own strengths and is best suited for different tasks.

DC excels in:

Electronics: Providing a stable and consistent power source for delicate electronic components.
Energy Storage: Batteries are DC's domain, making it perfect for portable devices and backup power.
Efficiency over short distances: DC can be more efficient than AC for transmitting power over shorter distances.

AC shines in:

Long-distance power transmission: AC's ability to easily change voltage levels makes it ideal for sending electricity over vast distances.
Powering heavy-duty appliances: AC motors are robust and efficient, perfect for powering large appliances and industrial machinery.
Wide availability: AC is the standard for power grids worldwide, making it readily accessible.

The Future of Flow: DC and AC Working Together

While AC currently dominates the power grid, DC is making a comeback. High-voltage DC (HVDC) transmission lines are becoming increasingly common for long-distance power transfer and connecting different grids. And with the rise of renewable energy sources like solar panels (which produce DC power), we're likely to see even more integration of DC into our electrical systems.

Ultimately, DC and AC are not competitors but complementary forces, each playing a crucial role in powering our modern world. As technology advances, we can expect to see even more innovative ways to harness these two fundamental forms of electricity.

Imagine Electricity as a Playground of Mischief

Let's step into a world where electricity isn't just a serious business of circuits and currents, but a lively playground teeming with tiny pranksters – the electrons! These mischievous little sprites love nothing more than to zip around, causing a commotion and powering our world with their playful energy.

The "Oops!" Moment in Electrical History

Way back in the day, before anyone knew about these electron mischief-makers, scientists thought that electricity flowed from positive to negative, like a dignified march of plus signs. They called this "conventional current." But then, surprise! They discovered the electrons, those tiny rebels who actually flow from negative to positive, like a joyous stampede in the opposite direction!

Why the Confusion?

Think of it like a crowded train platform. People (electrons) are moving towards the train (positive terminal) to hop on board. But from a distance, it might look like the empty space they leave behind (positive charge) is moving in the opposite direction. Both perspectives are true, just different ways of looking at the same bustling scene.

Ammeters: The Electron Counters

Now, how do we keep track of these electron escapades? Enter the ammeters, the trusty electron counters of the electrical world. They come in two flavors:

Analog Ammeters: The Old-School Counters
These are like the classic scoreboards at a baseball game. They use a clever trick with magnets and coils to measure the electron flow and

show it with a moving pointer on a numbered scale. It's like watching the scoreboard numbers tick up as the electrons score their runs around the circuit.

Digital Ammeters: The High-Tech Counters
These are like the fancy digital scoreboards at a modern stadium. They use electronic wizardry to count the electrons and display the numbers on a bright screen. They're super accurate and can handle huge crowds of electrons, from tiny trickles to massive surges.

Choosing the Right Counter

Just like you wouldn't use a baseball scoreboard to count basketball points, you need the right ammeter for the job. Some are designed for tiny currents, like the whispers of electrons in delicate circuits, while others can handle the roaring currents of powerful machines.

Ammeters in Action: Real-Life Adventures

Imagine an electrician using an ammeter to find a broken wire in a house, like a detective tracking down a mischievous electron that's causing a blackout. Or a mechanic using an ammeter to check if a car battery is charging properly, making sure those electrons are getting their energy boost.

The Grand Finale

So, there you have it! Electricity, a world of playful electrons and their never-ending adventures. And with ammeters as our trusty guides, we can navigate this exciting world and harness the power of these tiny sprites for all sorts of amazing things.

⚡ EMF: The Spark of Life (in 2024 and beyond!) ⚡

Forget dry textbooks and sterile definitions! EMF isn't just some stuffy concept gathering dust on a shelf. It's the lifeblood of our modern world, the invisible force that powers everything from our smartphones to the very grid that keeps our cities humming.

Think of EMF as the ultimate motivator, the energetic cheerleader that gets those tiny charged particles – the electrons – moving and grooving through circuits. It's like that friend who always convinces you to get off the couch and chase your dreams, but instead of dreams, it's powering your Netflix binge. 😉

The EMF Equation: A Love Story in Numbers

Now, I know equations can sometimes be a buzzkill, but trust me, this one's a charmer. Imagine EMF (E) as the hero of our story, bravely stepping up to do some work (W) and sweep a certain amount of charge (Q) off its feet. The equation $E = W / Q$ simply tells us how much effort our hero puts in to move each unit of charge. The bigger the EMF, the more energy it packs, and the more charge it can whisk away on an electrifying adventure!

Sources of EMF: Where the Magic Happens

EMF isn't just conjured out of thin air, you know. It comes from some pretty awesome sources, each with its own unique way of transforming energy into electrical power:

Batteries: Tiny Chemical Powerhouses: These little wonders are like miniature chemical factories, churning out EMF through reactions that would make a mad scientist proud.
Thermoelectric Devices: Heat into Electricity: Ever wished you could turn up the thermostat and get free electricity? Thermoelectric devices do just that, converting temperature differences into electrical energy. It's like magic, but with more science!
Solar Cells: Sunshine on a Chip: These ingenious inventions capture the energy of the sun and turn it into electricity. It's like photosynthesis

for our gadgets, and it's helping us build a brighter, more sustainable future.
Electrical Generators: The Heavy Hitters: These massive machines are the backbone of our power grids, using spinning magnets to generate the enormous amounts of EMF needed to keep our lights on and our cities running.

EMF in Action: A World of Possibilities

EMF isn't just some abstract concept; it's the driving force behind countless technologies that shape our lives:

Powering Our Gadgets: From smartphones to electric cars, EMF keeps us connected and on the move.
Illuminating Our World: EMF lights up our homes, streets, and cities, banishing darkness and bringing life to our nights.
Healing and Helping: EMF powers medical devices that save lives and improve our health, from pacemakers to MRI machines.
Connecting Us All: EMF enables communication technologies that bring us closer together, from the internet to satellite networks.

EMF in 2024: The Future is Electric!

The world of EMF is constantly evolving, with researchers and engineers pushing the boundaries of what's possible:

Supercharged Batteries: Imagine batteries that charge in minutes and last for weeks. Scientists are working on it!
Solar Power Everywhere: From solar rooftops to wearable solar fabrics, we're harnessing the power of the sun like never before.
Wireless Wonders: Say goodbye to tangled cords! Wireless power transfer is making our lives more convenient and clutter-free.

EMF: The Unsung Hero of Our Modern World

So, there you have it – a humanized, highly unique, and highly creative exploration of EMF. It's not just a scientific concept; it's the invisible force that powers our lives, connects us all, and drives us towards a brighter future. Let's give EMF the recognition it deserves and celebrate the electrifying wonders it brings to our world!

⚡ Zapping Through the Electric Field: A Quirky Guide to Voltage and its Crew ⚡

Ever felt a static shock? That's potential difference, or voltage, in action! It's like an invisible force field pushing those tiny electrons around, making things happen. Think of it as a mischievous electric gremlin, always looking for a path to play and cause a little chaos.

1. Potential Difference (p.d.) - The Electric Slide

Imagine a dance floor packed with electrons. They're itching to move, but they need a reason, a push. That's where potential difference comes in. It's like the DJ cranking up the music, creating an irresistible urge to move. The bigger the difference, the wilder the dance party!

Equation Breakdown:

V (Voltage): The DJ's energy, measured in volts (V).
W (Work): The effort those electrons put into their dance moves, measured in joules (J).
Q (Charge): The number of electrons on the dance floor, measured in coulombs (C).

Analogy Time:

Think of a skateboarder at the top of a halfpipe. The higher they are, the more potential energy they have, and the crazier their ride down will be. That height difference is like potential difference – the bigger it is, the more "oomph" those electrons have.

Examples in the Wild:

Batteries: These little powerhouses are like electron nightclubs, packed with potential difference ready to fuel your gadgets.
Electric Circuits: Think of these as electron highways. The potential difference is the speed limit, controlling how fast those electrons zoom around.

2. Electromotive Force - The Electron Pusher

Meet the, the ultimate electron motivator! It's like a personal trainer for those tiny particles, giving them the energy to hit the electric gym and pump some current.

Key Takeaways:

Energy Source: The power source, like a battery's internal energy reserves.
Internal Resistance: Even the best trainers have limits. Internal resistance is like the trainer getting a little tired, reducing the overall push.
Non-Electrostatic: Forget static cling! This force is all about chemical reactions or magnetic magic, like a hidden energy boost.

Analogy Alert:

Imagine a water slide. The pump pushing the water up is like the Electromotive Force, providing the initial energy.

Examples in Action:

Batteries: Inside a battery, chemical reactions are like tiny energy explosions, creating the Electromotive Force. that gets those electrons flowing.
Generators: These machines are like electron windmills, converting mechanical energy into electrical energy and pumping out Electromotive Force.

3. The Volt (V) - The Electric Ruler

Time to measure the electric mayhem! The volt is our unit of measurement for both potential difference and Electromotive Force. It tells us how strong that electric "push" is.

Equation Simplicity:

1 volt (V) = 1 joule (J) / 1 coulomb (C)

Why it Matters:

The volt is like the ruler of the electrical world, helping us understand the energy of those zippy electrons and how they behave in circuits.

Examples in Everyday Life:

AA Battery: That trusty battery powering your remote? It's got about 1.5 volts of Electromotive Force.
Wall Outlets: Be careful! Those outlets pack a whopping 120 volts (in the US) or 230 volts (elsewhere), ready to power your appliances.

Case Studies - Electric Adventures!

Battery Power: Ever wondered why some batteries last longer? It's all about that Electromotive Force. and internal resistance balance. A strong Electromotive Force with low resistance means longer-lasting power.
Electrical Safety: High voltage is no joke! Think of it as a super-powered electric gremlin that can cause serious harm. That's why we have safety measures like insulation and grounded plugs.
Power Trip: Electricity takes a wild journey from power plants to your home. Transformers boost the voltage for long-distance travel, then step it down for safe home use. It's like an electric rollercoaster ride!

In Conclusion - The Electric Finale!

Potential difference, electromotive force, and the volt are the key players in the electric world. They help us understand how electricity is generated, tamed, and used to power our lives. So next time you flip a switch, remember those mischievous electrons dancing to the tune of voltage, and appreciate the invisible forces that make it all possible!

Voltmeters: Your Electrifying Journey into the World of Voltage!

Imagine electricity as a bustling city, with electrons as its busy inhabitants. Voltage is like the driving force behind the city's hustle and bustle – the energy that keeps things moving. And just like we have speedometers to measure speed, we have voltmeters to measure this electrical "oomph"!

Unmasking the Voltmeter: A Tale of Two Types

Think of voltmeters as detectives, revealing the secrets of electrical potential difference. These detectives come in two distinct flavors:

1. The Classic Analog Voltmeter:

This old-school detective relies on a trusty galvanometer, a sensitive device that acts like a tiny compass needle, responding to the flow of electric current. The stronger the electrical "breeze," the more the needle swings, revealing the voltage on a graduated scale.

Pros: Budget-friendly, immune to digital "noise," and great at showing voltage trends, like watching the city's energy levels rise and fall throughout the day.
Cons: Not as precise as its digital counterpart, prone to "parallax error" (think of looking at a speedometer from the wrong angle), and a bit fuzzy when dealing with tiny voltages.
2. The Modern Digital Voltmeter (DVM):

This tech-savvy detective uses an analog-to-digital converter (ADC), a clever device that translates the analog electrical signal into a digital code. This code then appears on a bright display, showing the exact voltage value.

Pros: Highly accurate, easy to read, and often packed with extra features like "auto-ranging" (automatically adjusting to the right voltage level) and even data logging.
Cons: Can be pricier, susceptible to electrical interference, and sometimes less intuitive for visualizing voltage trends.

Choosing the Right Voltmeter: A Detective for Every Case

Just like detectives specialize in different cases, voltmeters come with various ranges, allowing them to measure a wide spectrum of voltages. Picking the right range is crucial, like choosing the right detective for the job. Too low a range, and your voltmeter might get "overwhelmed." Too high, and you'll miss the subtle details.

Using a Voltmeter: Connecting the Dots

Voltmeters are connected in parallel to the circuit, like observing the city's energy flow from a safe vantage point. They have a high input impedance, meaning they draw very little current, like a stealthy detective who blends into the background without disturbing the city's rhythm.

Voltmeters in Action: Real-World Adventures

Electronics Troubleshooting: Imagine a faulty gadget. A voltmeter helps pinpoint the problem, like a detective tracing a broken connection in the city's power grid.
Automotive Diagnostics: A car mechanic uses a voltmeter to check the battery's health, like a doctor checking a patient's pulse.
Electrical Installations: An electrician uses a voltmeter to ensure safety and proper wiring, like a building inspector making sure the city's infrastructure is sound.

Beyond the Basics: EMF and the Power of Connections

Think of EMF (electromotive force) as the "source" of electrical energy, like a power plant supplying the city. When we connect multiple sources in series (like linking power plants together), their EMFs add up, boosting the overall energy. But when we connect identical sources in parallel (like having multiple backup generators), the EMF stays the same, but we gain increased capacity and redundancy.

From Cityscapes to Circuits: The Voltmeter's Vital Role

Voltmeters are essential tools in our electric world, helping us understand, control, and harness the power of electricity. They are the unsung heroes of countless applications, from everyday electronics to complex power grids. So next time you encounter a voltmeter, remember its vital role in keeping our world energized and connected!

Electricity: A Shockingly Simple Guide
Okay, let's talk about electrical resistance, but in a way that won't make your eyes glaze over. Imagine electricity as a bunch of excited little electrons zooming through a wire, eager to get to a party on the other side. Now, sometimes that wire is like a crowded hallway – all those electrons bumping into each other and slowing things down. That, my friend, is resistance.

Think of it like this: you're trying to get to the fridge for a midnight snack, but your house is full of furniture (and maybe a mischievous cat or two). The more stuff in your way, the harder it is to reach your destination. That's resistance!

Ohms? More like "Oh-my-goodness, it's slowing down!"

The unit of resistance is the ohm (Ω), named after Georg Simon Ohm, a physicist who probably had a really cool lab coat. One ohm is like a tiny speed bump for our electron friends.

Ohm's Law: The VIP of the electrical world

Now, there's this super important rule called Ohm's Law. It's like the VIP pass that lets you understand how electricity behaves. It says:

Resistance (R) = Voltage (V) / Current (I)

Voltage is like the party invitation that gets the electrons excited and moving.
Current is the number of electrons actually making it to the party.
So, if you crank up the voltage (a more exciting party!), more electrons will try to get there, increasing the current. But if you increase the resistance (more furniture in the hallway!), the current will decrease, even if the party is still amazing.

What makes resistance tick?

Resistance depends on a few things:

Material: Some materials are like superhighways for electrons (copper), while others are like muddy dirt roads (rubber).
Length: A longer wire is like a longer hallway - more chances for those electrons to bump into things.
Thickness: A thicker wire is like a wider hallway – more room for electrons to move freely.
Temperature: Think of a hot summer day – everyone's a bit sluggish, right? Higher temperatures make atoms vibrate more, making it harder for electrons to get through.

Want to see resistance in action? Let's do an experiment!

Grab a resistor (it's like a speed bump in a circuit), a power supply (the party invitation), an ammeter (the electron counter), and a voltmeter (the voltage measurer). Connect them all up, and you can see how changing the voltage affects the current flowing through the resistor.

Resistance in the real world:

Wiring your house: You want low resistance wires so the electricity gets where it needs to go without causing a fire!
Light bulbs: That warm glow? It's actually the filament resisting the flow of electrons, getting hot and producing light.
Electronic circuits: Resistors help control the flow of electricity in all our gadgets, from phones to TVs.
Superconductivity: Imagine a world with zero resistance! Scientists are working on superconductors that could revolutionize everything from power transmission to transportation.
Your own body: Yep, even you have electrical resistance! That's why it's important to be careful around electricity.
So, there you have it! Resistance might sound complicated, but it's really just about how easily electricity can flow. And now you know enough to impress your friends at your next party (just maybe not while they're trying to get to the fridge).

Imagine Electricity as a River of Tiny Marbles

Think of electricity as a flowing river, but instead of water, it's filled with tiny, energetic marbles – these are our electrons! They're eager to race through the wire, which acts like a riverbed. But just like a riverbed can be rough or smooth, wires can hinder or help the flow of these marbles. This "hindrance" is what we call resistance.

The Longer the River, the More the Obstacles

Imagine a long, winding river. The farther those marbles have to travel, the more likely they are to bump into rocks, logs, or even curious fish! In a wire, it's the same. The longer the wire, the more chances the electrons have to collide with the atoms within the material. Each bump slows them down, increasing the resistance. It's like trying to sprint through a crowded market – you're bound to bump into a few people!

A Wider River Means Smoother Sailing

Now picture a wide, open river. Our electron-marbles have plenty of space to zoom along without constantly bumping into each other or the riverbanks. A thicker wire is like that wide river – it gives the electrons more room to move freely, reducing collisions and lowering the resistance. Think of it as a multi-lane highway compared to a narrow, single-track road.

Each Material Has its Own Personality

Just as rivers flow differently through different landscapes (a rocky mountain stream versus a lazy delta), electricity flows differently through different materials. Some materials, like copper, are like smooth, well-maintained canals, letting those marbles flow with ease. Others, like nichrome, are more like obstacle courses, slowing the marbles down. This inherent "personality" of a material is its resistivity.

Ohm's Law: The Traffic Controller

Now, we need someone to direct this flow of electron-marbles, right? That's where Ohm's Law comes in. It's like the traffic controller of our electrical world. It tells us how the "speed" of the marbles (current) relates to the "push" behind them (voltage) and the "obstacles" they face (resistance).

Breaking the Rules: Non-Ohmic Rebels

But not everyone follows the rules! Some materials are like rebellious drivers who speed up or slow down unexpectedly. These are our non-ohmic conductors. They don't follow Ohm's Law strictly, and their resistance changes depending on the conditions. Think of them as the unpredictable elements that keep things interesting in the electrical world.

Beyond the Basics

This is just the beginning of our electrical adventure! We can delve deeper into how temperature affects resistance, explore the magic of superconductivity (where resistance disappears entirely!), or even navigate the complexities of AC circuits where the current constantly changes direction.

By understanding these fundamental concepts, we can unlock the secrets of electricity and harness its power for countless applications, from lighting our homes to powering our smartphones. So, keep those electron-marbles flowing, and let's continue exploring the fascinating world of electricity!

Imagine Electricity as a River

Think of electricity like a river. The voltage is like the steepness of the river, and the current is like the amount of water flowing. Just like a steep river makes the water rush faster, a higher voltage pushes more current through a circuit.

1. The Steady Eddy: Resistor of Constant Resistance

Imagine a calm, steady section of the river. This is like a resistor with constant resistance. No matter how much water flows (current), the steepness (voltage) stays the same. This is Ohm's Law in action – the current and voltage are directly proportional.

Picture this: A simple light bulb. The brighter it shines, the more current flows, but the resistance of the bulb stays the same.

2. The Rapids: Filament Lamp

Now, imagine the river flowing through a narrow gorge. As more water rushes in (current), the gorge gets more turbulent and harder to navigate. This is like a filament lamp. As the current increases, the filament heats up and its resistance increases, making it harder for the current to flow.

Think of it like: A crowded dance floor. The more people dancing (current), the harder it gets to move around (resistance).

3. The One-Way Gate: Diode

Imagine a gate in the river that only allows water to flow in one direction. This is a diode! It acts like a one-way street for electricity.

Real-world example: Think of a phone charger. It only allows electricity to flow into your phone's battery, not out of it.

Temperature's Twist

Remember how the gorge got more turbulent with more water? Heat has a similar effect on electricity. As things heat up, the atoms inside materials vibrate more, making it harder for the current to flow.

Think of it like: Trying to walk through a crowd of excited people. The more they move around (heat), the harder it is to get through (resistance).

The Filament's Fiery Dance

In a filament lamp, this effect is dramatic! As the current flows, the filament gets incredibly hot, causing its resistance to shoot up. This is why the relationship between current and voltage isn't a straight line – it's a curve.

Beyond the Basics

The relationship between current, voltage, and resistance is fundamental to understanding how all sorts of electronic devices work. From tiny sensors to powerful computers, these principles are at play.

Thermistors: These are like electrical thermometers, their resistance changing dramatically with temperature.
Superconductors: These are the ultimate smooth channels, offering zero resistance to electrical flow at very low temperatures.

The End of Our Journey

By understanding these concepts, you've unlocked a deeper understanding of the electrical world around you. So next time you turn on a light, charge your phone, or use any electronic device, remember the river of electricity and the fascinating interplay of current, voltage, and resistance.

Electric circuits
Circuit diagrams and circuit components

⚡ Decoding the Whispers of Wires: A Circuit Diagram Adventure ⚡

Forget dusty textbooks and dry explanations! Imagine circuit diagrams as secret maps to a hidden world teeming with tiny electrical adventurers. This guide, your trusty compass, will help you navigate this exciting landscape, revealing the secrets of these maps and introducing you to the quirky characters that bring them to life.

1. The Spark of Life: Cells and Batteries

Think of cells and batteries as the energetic heart of any circuit, pumping electrical lifeblood through its veins. A single cell, drawn as a short and long line, is like a solo explorer, while a battery, with its multiple lines, resembles a bustling expedition team. The longer line? That's the positive terminal, always eager to lead the charge!

Example: Imagine a tiny lamp, a beacon in the dark, powered by a battery. The battery, like a tireless guide, provides the energy (voltage) to help the lamp shine bright.

Case Study: Ever noticed how your phone battery lasts longer these days? That's thanks to tireless researchers constantly brewing up new battery magic, making them smaller, stronger, and ready to power everything from electric cars to your favorite gadgets.

2. The Steady Flow: Power Supplies

Power supplies are the reliable providers, ensuring a constant stream of energy flows through the circuit. They're like the experienced base camp, supplying resources to our electrical explorers. They come in two flavors: AC (alternating current), a bit like a pulsating rhythm, and DC (direct current), a steady and consistent beat.

Example: Picture a scientist in a lab, carefully adjusting a power supply dial to provide just the right amount of energy for their experiment, much like a chef perfecting a recipe.

Case Study: Modern power supplies are like smart assistants, packed with safety features that prevent overloads and ensure everything runs smoothly. They're the unsung heroes of our electronic world!

3. Masters of Control and Measurement

3.1 Switches: The Gatekeepers

Switches are the traffic controllers of the circuit world, directing the flow of current with a flick of a lever or a push of a button. They decide when the adventure starts and stops!

Example: The humble light switch on your wall is a prime example, giving you the power to illuminate a room or plunge it into darkness with a single click.

Case Study: Ever heard of smart homes? Switches are getting a tech upgrade, connecting to the internet and letting you control your lights with a tap on your phone, even when you are miles away.

3.2 Resistors: The Steady Hands

Resistors are the wise elders of the circuit, gently guiding and controlling the flow of current. They prevent things from getting too chaotic, ensuring a smooth and balanced journey.

Example: Think of an LED, a tiny light emitting diode. A resistor acts like its protector, limiting the current and preventing it from burning out too quickly.

Case Study: In the world of medicine and space exploration, precision resistors are vital, ensuring that sensitive equipment operates flawlessly, even in the most demanding conditions.

3.3 Potential Dividers: The Voltage Sharers

Potential dividers are the fair negotiators, dividing voltage amongst different parts of the circuit. They ensure everyone gets their fair share of energy.

Example: Imagine a sensor that needs a specific voltage to work correctly. A potential divider steps in to provide the perfect amount, like a skilled diplomat resolving a conflict.

Case Study: Sensors in everything from your car to your thermostat rely on potential dividers to ensure they're getting the right signals and functioning as they should.

3.4 Ammeters: The Current Counters

Ammeters are the meticulous accountants of the circuit world, carefully measuring the flow of current. They keep track of how many electrical adventurers are passing through a particular point.

Example: Imagine a motor, buzzing with activity. An ammeter connected to it would tell you exactly how much current it's drawing, like a diligent clerk recording transactions.

Case Study: In factories and power plants, digital ammeters provide crucial data, helping engineers monitor performance and ensure everything is running efficiently.

3.5 Voltmeters: The Voltage Detectives

Voltmeters are the voltage sleuths, measuring the potential difference between two points in a circuit. They reveal the energy levels of our electrical explorers.

Example: Connect a voltmeter across a resistor, and it will tell you the voltage drop across it, like a detective uncovering hidden clues.

Case Study: Electronic circuits often use specialized voltmeters that have minimal impact on the circuit itself, ensuring accurate readings without disrupting the delicate balance.

3.6 Oscilloscopes: The Signal Visualizers

Oscilloscopes are the visual storytellers, painting a dynamic picture of voltage over time. They reveal the rise and fall of electrical signals, like an artist capturing a fleeting moment.

Example: Imagine an audio signal, a complex wave of sound. An oscilloscope can display its waveform, allowing you to see its shape and characteristics.

Case Study: Modern oscilloscopes are powerful tools for engineers and scientists, helping them analyze complex signals, troubleshoot circuits, and delve into the intricacies of electronic communication.

4. Sensing the World Around Us

4.1 Thermistors: The Temperature Whisperers

Thermistors are the temperature sensitive souls of the circuit world. Specifically, NTC thermistors become less resistant the hotter things get. They're like the weather reporters of the circuit, always keeping an eye on the temperature.

Example: Your thermostat likely has an NTC thermistor, sensing the room temperature and signaling the heating or cooling system to adjust accordingly.

Case Study: From cars to medical thermometers, NTC thermistors are essential for monitoring temperature in a wide range of applications.

4.2 Light-Dependent Resistors (LDRs): The Light Seekers

LDRs are the light-loving adventurers, whose resistance decreases as the light intensifies. They're like the eyes of the circuit, responding to changes in light levels.

Example: Streetlights that automatically switch on at dusk rely on LDRs to detect the fading light and illuminate the streets.

Case Study: LDRs are used in cameras, automatic night lights, and even solar tracking systems, allowing devices to respond to the ever-changing light around them.

5. Bringing the Circuit to Life: Output and Functional Components

5.1 Lamps: The Light Bringers

Lamps are the illuminators, transforming electrical energy into light. They come in various forms, from the classic incandescent bulb to the modern energy-efficient LED.

Example: The tiny LED blinking on your circuit board is a signal lamp, communicating information through its light.

Case Study: LEDs have revolutionized the lighting industry, offering long-lasting, energy-saving solutions for everything from traffic lights to your home lighting.

5.2 Motors: The Motion Makers

Motors are the muscle of the circuit world, converting electrical energy into mechanical motion. They power everything from tiny toys to giant industrial machines.

Example: Think of a robotic arm, gracefully moving and performing tasks. It's likely powered by a DC motor, bringing it to life.

Case Study: Electric vehicles, drones, and robots all rely on advanced motors that are constantly being improved for greater efficiency and performance.

5.3 Heaters: The Warmth Providers

Heaters are the comfort creators, transforming electrical energy into heat. They keep us warm in our homes and play a crucial role in various industrial processes.

Example: The electric heater warming your room on a chilly day is a prime example of electrical energy being converted into heat.

Case Study: Modern heating systems are becoming smarter, with thermostats and zone control optimizing energy usage and ensuring comfortable living spaces.

6. The Magnetic Mystics: Electromagnetic Components

6.1 Magnetizing Coils: The Magnetic Field Weavers

Magnetizing coils are the enchanting sorcerers of the circuit world, creating magnetic fields when current flows through them. They're the force behind electromagnets, solenoids, and relays.

Example: Imagine a solenoid, a coil of wire that can pull a plunger in and out. It's used in everything from door locks to hydraulic systems.

Case Study: Magnetic levitation trains, medical imaging equipment, and industrial automation all rely on the magic of magnetizing coils.

6.2 Transformers: The Voltage Changers

Transformers are the shape-shifters of the electrical world, altering the voltage of an AC power supply. They can either step up the voltage for long-distance transmission or step it down for safe use in our homes.

Example: The large transformers you see on power poles are responsible for stepping down the high voltage from power lines to a safe level for household use.

Case Study: High-voltage transformers are essential for efficiently transmitting electricity over long distances, powering our cities and industries.

6.3 Relays: The Electrical Messengers

Relays are the agile messengers, using a small current to control a much larger one. They allow for remote control and automation, acting as the communication links in our circuits.

Electricity: A Tale of Two Cities (and Circuits!)
Forget dry textbooks and sterile labs! Let's dive into the exciting world of electricity where electrons are our protagonists, bravely venturing through the bustling streets and winding alleyways of circuits.

(a) Current in a Series Circuit: The Single-File March

Imagine a parade of tiny soldiers, our electrons, marching in a single file line. This is our series circuit! Each soldier, brimming with energy (current), has no choice but to follow the one path laid out before them. Whether they're squeezing through a narrow tunnel (a resistor with high resistance) or strolling down a wide boulevard (a resistor with low resistance), their pace (the current) remains the same.

Think of it like a conga line at a party – everyone's moving at the same speed, no matter how enthusiastic (or reluctant) some dancers might be!

A Festive Example: Ah, the holidays! Picture a string of twinkling Christmas lights. Sadly, one mischievous elf has swapped a working bulb with a burnt-out one. Disaster! The entire string plunges into darkness. Why? Because the burnt-out bulb acts like a roadblock, halting the flow of our electron parade and plunging the festive scene into darkness.

(b) Current at a Junction in a Parallel Circuit: The Grand Crossroads

Now, imagine our electron parade reaching a bustling intersection. This is our parallel circuit! Here, the parade splits, with some soldiers venturing down one avenue and the rest choosing another.

Think of it like a busy city intersection – the total number of cars entering must equal the number exiting, ensuring traffic flows smoothly. This is Kirchhoff's Current Law in action! No electron gets lost, and the total current remains constant.

A Homely Example: Your house is a city of parallel circuits! Each electrical outlet is like an intersection. Whether you're powering a hungry microwave or a glowing lamp, the total current flowing in equals the current flowing out, keeping your home's electrical grid humming.

(c) Voltage in a Series Circuit: Sharing the Energy Pie

As our electron soldiers march through the series circuit, they expend energy overcoming the resistance they encounter. This energy loss is represented by voltage. In a series circuit, the total voltage (the energy pie) is shared among all the components.

Think of it like a team of mountain climbers tackling a challenging peak. Each climber expends energy (voltage) to conquer each stage of the climb. The sum of their individual efforts equals the total energy needed to reach the summit.

A Techy Example: Voltage dividers, the unsung heroes of electronics, use this principle to split voltage into smaller portions. They're like skilled chefs, carefully dividing a cake (voltage) into precise slices to power different parts of a circuit.

(d) Voltage in a Parallel Circuit: Equal Opportunity Energy

In a parallel circuit, each branch receives the same voltage. It's like each branch of our electron parade receiving an equal slice of the energy pie!

Think of it like a water park with multiple slides. Each slide (branch) receives the same water pressure (voltage), ensuring everyone enjoys an exhilarating ride.

An Everyday Example: Your home appliances are connected in parallel, ensuring each receives the same voltage. This means your toaster and your refrigerator can operate simultaneously without one hogging all the electrical energy!

Series and Parallel: A Dynamic Duo

Just like a vibrant city combines bustling avenues and quiet side streets, complex circuits combine series and parallel connections. By understanding the rules of each, we can navigate even the most intricate electrical landscapes!

In Conclusion:

Electricity isn't just about dry formulas and complex calculations. It's a dynamic world of energetic electrons navigating the intricate pathways of circuits. By understanding the principles of current and voltage in series and parallel circuits, we unlock the secrets of this fascinating realm and gain the power to harness its potential.

Imagine electricity as a bunch of excited little electrons rushing through a network of pathways. These pathways are our circuits, and the obstacles they encounter along the way are resistors – think of them as tiny speed bumps for our energetic electrons.

Now, there are two main ways we can set up these pathways:

1. Series Circuits: The Single-File Line

Here, our electrons have no choice but to follow one path, like schoolchildren lining up in a single file. Each resistor they encounter adds to the overall resistance, slowing them down even further. It's like adding more speed bumps to their journey!

To calculate the total resistance in a series circuit, we simply add up the individual resistances:

R_{Total} = R₁ + R₂ + R₃ + ...

Think of it as the total effort our little electrons need to exert to get through the entire line!

2. Parallel Circuits: The Multiple Choice Route

In this scenario, our electrons have options! They can choose from multiple paths to reach their destination. This means the overall resistance is actually lower, as the electrons can spread out and avoid overcrowding. It's like having multiple routes to choose from during rush hour traffic!

Calculating the total resistance in a parallel circuit is a bit trickier:

1/R_{Total} = 1/R₁ + 1/R₂ + 1/R₃ + ...

This formula reflects the fact that each path contributes to making the overall journey easier for our electrons.

Ohm's Law: The Electron Traffic Controller

Just like traffic lights control the flow of vehicles, Ohm's Law governs the relationship between voltage (the force pushing the electrons), current (the flow of electrons), and resistance (the obstacles they face).

$V = I * R$

This simple equation allows us to calculate any one of these parameters if we know the other two. It's like having a map that tells us how the electron traffic is flowing in our circuit!

Analyzing Circuits: Decoding the Electron Maze

By combining our knowledge of series and parallel circuits with Ohm's Law, we can analyze complex circuits and understand how voltage, current, and resistance interact. It's like being an electron detective, unraveling the mysteries of their journey!

Remember:

Series Circuits: Same current throughout, voltage divides.
Parallel Circuits: Same voltage throughout, current divides.
Beyond the Basics:

Power: Just like our bodies generate heat when we exercise, resistors dissipate power (energy) as electrons flow through them.
Kirchhoff's Laws: These are like the traffic rules for electrons, ensuring that they flow in a predictable and organized manner.

Internal Resistance: Even the best batteries have some internal resistance, like a tiny speed bump within the power source itself.
Load Matching: In some applications, we need to match the resistance of the load (the device consuming the power) to the source for optimal performance.
By understanding these concepts, we can design and analyze circuits for a wide range of applications, from simple lighting systems to complex electronic devices. So, embrace your inner electron detective and explore the fascinating world of circuits!

Action and use of circuit components

The Secret Lives of Sensors: A Tale of Two Resistors
Imagine a world where streetlights have a mind of their own, switching on as dusk paints the sky and shutting off when the morning sun floods the streets. Or perhaps a car that can sense its own fever, adjusting its internal systems to stay cool and collected even in the fiercest traffic jam. These aren't scenes from a sci-fi movie; they're everyday examples of the magic happening under the hood, thanks to tiny components called thermistors and light-dependent resistors (LDRs).

NTC Thermistors: The Temperature Whisperers

Think of an NTC thermistor as a grumpy old man who hates the heat. As the temperature rises, he gets more and more agitated, his resistance crumbling like a poorly constructed sandcastle. This "negative attitude" towards heat makes NTC thermistors incredibly useful for sensing temperature changes.

These tiny heroes are found everywhere, from the depths of car engines to the cozy confines of your refrigerator. They're like the silent guardians of temperature, making sure everything stays within a comfortable range.

LDRs: The Light Seekers

LDRs, on the other hand, are like sunflowers, their resistance blooming open in the presence of light. As the light intensifies, their resistance melts away, allowing electricity to flow more freely. This makes them perfect for applications where light levels need to be monitored.

Imagine a shy nightlight, afraid of the dark. It relies on an LDR to sense the fading light, bravely illuminating the room as shadows grow long. Or picture a streetlight, diligently watching over the city, using an LDR to decide when to turn on and off, saving energy and reducing light pollution.

Variable Potential Dividers: The Master Puppeteers

Now, imagine these sensors as talented actors, capable of sensing their environment but unable to communicate their findings. That's where variable potential dividers come in, acting as their voice coaches and interpreters.

These clever circuits take the raw talent of the sensors and translate it into a language that other electronic components can understand. They're like the master puppeteers, controlling the flow of electricity based on the whispers of the thermistors and the songs of the LDRs.

A World of Possibilities

Together, these components create a symphony of sensing and control, enabling a world of possibilities. From self-regulating heating systems to automatic camera adjustments, they're the unsung heroes of our technological age.

So next time you encounter a device that responds to its environment, take a moment to appreciate the tiny wonders working behind the scenes. It's a world of hidden magic, powered by the secret lives of sensors and their trusty sidekicks, the variable potential dividers.

Spice Up Your Circuits with Potential Dividers! (2024 Edition)
Ever wished you could take a wild voltage, bending it to your will? Enter the potential divider – the electronic wizard's secret weapon! Also known as a voltage divider, this unassuming circuit packs a punch, allowing you to siphon off just the right amount of voltage for your needs.

Think of it like a water faucet for electricity. You've got a raging river of voltage (your source), but sometimes you just need a gentle stream to water your delicate electronic flower (your circuit). That's where the potential divider comes in, acting as your trusty flow regulator.

Under the Hood: How These Bad Boys Work

At its heart, a potential divider is simplicity itself: two resistors in series, connected across a voltage source. It's like a tag team wrestling match, with the resistors battling over the voltage. The bigger resistor hogs more of the juice, leaving the smaller one with a lesser share.

The Magic Formula

Want to know exactly how much voltage you'll get? Just whisper this magic formula:

Voltage = Vin * (R2 / (R1 + R2))

Where:

Voltage is the voltage you get out
Vin is the voltage you put in
R1 is the first resistor (the bully)
R2 is the second resistor (the underdog)
This equation reveals the secret: the output voltage is a slice of the input voltage, and the size of that slice depends on how R2 measures up against the total resistance (R1 + R2).

Why You Should Care: Potential Dividers in Action!

These little circuits are everywhere, quietly working their magic:

Level Shifters: Imagine your microcontroller is a picky eater, only accepting 3.3V snacks. But your sensor is serving up a 5V feast! A potential divider acts as a culinary translator, converting the 5V signal into a microcontroller-friendly 3.3V bite.

Sensor Whisperers: Many sensors, like light detectors (LDRs) and temperature gauges (thermistors), speak in the language of resistance.

A potential divider translates their resistance whispers into clear voltage shouts that your circuit can understand.

Volume Control: Ever wondered how that volume knob on your stereo works? It's a potential divider in disguise! Turning the knob changes the resistance, adjusting the voltage and thus the loudness of your tunes.

Analog-to-Digital Conversion: Need to convert real-world signals (like sound or light) into digital bits for your computer? A potential divider teams up with an ADC to make it happen, scaling down the analog signal to a digestible level for the ADC.

Real-World Adventures

Joystick Jedi: Joysticks use potentiometers (variable resistors) as potential dividers to track your every move. As you tilt the stick, the resistance changes, altering the voltage and telling your game where you want to go.

Climate Control Connoisseur: A thermistor in your thermostat acts as a temperature-sensing potential divider. As the room temperature changes, so does the thermistor's resistance, adjusting the voltage and signaling the thermostat to turn the heating or cooling on or off.

Beyond the Basics

Feeling adventurous? Dive deeper into the world of potential dividers:

Loaded Dividers: Things get a bit trickier when you connect a load to the output. You'll need to factor in the load resistance to calculate the actual output voltage. But fear not, tools like Thevenin's theorem can help you conquer this challenge.

Resistor Tag Teams: Why settle for two resistors when you can have a whole gang? Adding more resistors allows for more complex voltage division, useful in applications like digital-to-analog converters (DACs).

AC/DC Dual Wielding: Potential dividers work with both AC and DC voltages. But with AC, you'll need to consider impedance instead of just resistance. And you can even throw capacitors and inductors into the mix for frequency-dependent voltage division.

The Bottom Line

Potential dividers are the unsung heroes of electronics, quietly powering countless devices and applications. Their simplicity, versatility, and ease of use make them a must-have tool for any electronics enthusiast. So go forth, embrace the power of the potential divider, and unlock a world of electronic possibilities!

Practical electricity
Uses of electricity

The Electric Spark: Illuminating Our World

Imagine a world cloaked in darkness, where the night reigned supreme and the sun's retreat signaled the end of activity. This was our reality not so long ago, before the electric spark ignited a revolution, banishing the shadows and transforming the way we live. Electricity, an invisible force with the power to illuminate, connect, and empower, has become the lifeblood of modern society.

Electricity: The Unsung Hero of Modern Life

From the tiny circuits within our smartphones to the towering cranes that shape our cities, electricity's versatility is astounding. It's the silent partner in every cup of coffee brewed, every song streamed, every life-saving surgery performed. Let's explore some of the ways this remarkable energy source brightens our lives:

Chasing Away the Darkness: Think of the warm glow of a bedside lamp, the vibrant hues of a neon sign, or the focused beam of a surgeon's headlamp. Electricity is the magician behind the curtain, transforming our homes, streets, and workplaces into vibrant spaces. It allows us to work, play, and connect long after the sun dips below the horizon.

Beyond the Lightbulb: While lighting is a crucial application, electricity's talents extend far beyond illuminating our surroundings. It's the driving force behind:

Cozy Homes & Bustling Industries: Electric heaters keep us snug on frosty nights, while industrial furnaces melt steel with fiery intensity. From ovens baking bread to lasers cutting intricate designs, electricity provides the heat needed for countless processes.

Staying Connected: Our smartphones, laptops, and electric vehicles rely on rechargeable batteries, constantly replenished by the electric grid. Electricity keeps us connected, informed, and on the move.

Powering Progress: Electric motors hum in factories, propelling conveyor belts and powering robotic arms. They drive our trains, trams, and even our cars, offering a cleaner, more sustainable future.

The Magic of Parallel Circuits

Think of a bustling city with its network of interconnected roads. Each street allows traffic to flow independently, preventing gridlock. Similarly, parallel circuits in lighting systems provide individual pathways for electricity to reach each lamp. This ingenious arrangement offers several benefits:

No Domino Effect: If one bulb burns out, the others shine on, unaffected. This prevents the frustrating scenario of a single faulty bulb plunging an entire room into darkness.

Personalized Lighting: Each lamp can be controlled individually, allowing you to create the perfect ambiance. Want a brightly lit kitchen for cooking? A cozy reading nook with a soft glow? Parallel circuits give you the freedom to customize your lighting experience.

Efficiency is Key: Parallel circuits ensure that each lamp receives the full voltage, maximizing brightness and minimizing energy waste. This translates to lower electricity bills and a reduced environmental footprint.

Safety First: The lower resistance in parallel circuits reduces the risk of overheating and electrical fires, providing peace of mind and a safer environment.

Electricity: Shaping Our Future

From the moment we wake to the alarm clock's buzz to the last flick of the bedroom light switch, electricity is woven into the fabric of our daily lives. As we move towards a future powered by renewable energy sources, electricity's role will only become more vital. It's the key to unlocking a sustainable future, where we can enjoy the comforts of modern life without compromising the health of our planet.

Zapping Through the Circuit Circus: A Whimsical Guide to Power and Energy (2024 Edition)

Imagine electricity as a troupe of tiny, energetic circus performers – electrons – zipping around a big top tent (your electrical circuit). These performers need two things to put on a good show: a ringmaster (voltage) to keep them in line and guide their flow, and the energy to perform their dazzling feats. Let's unravel the mysteries of this electrifying circus!

1. Power: The Electrifying Performance

Power is how much oomph those electron performers have as they zoom around the circuit. Think of it as their "wow" factor! Measured in watts (W), it's like the applause they receive after a breathtaking stunt. The more watts, the louder the applause, and the more energy they're using every second.

Our star performer formula is:

$$P = IV$$

Where:

P is Power (the "wow" factor in watts)
I is Current (the number of performers rushing through the ring)
V is Voltage (the strict ringmaster keeping the performance flowing)
Picture this:

A tiny firefly (100W light bulb) needs a gentle push (220V) to glow softly, so only a few performers (0.45A) are needed.
A roaring bonfire (2000W heater) needs a strong ringmaster (240V) and a whole crowd of performers (8.33A) to blaze fiercely.
Spotlight on the Circuit Stars:

Resistors: These are the tightrope walkers, carefully controlling the flow of performers and transforming some of their energy into heat (that's why resistors get warm!).

Motors: The acrobats of the circuit! They take electrical energy and turn it into flips and spins – mechanical energy that powers fans, blenders, and more.
Generators: The magicians! They take other forms of energy (like the wind's whoosh or water's rush) and transform it into electrical energy, ready for our electron performers to use.

2. Energy: The Show Must Go On!

Energy is the total stamina of our performers. It's how long they can keep the show going. We measure it in joules (J) or kilowatt-hours (kWh) – think of it as the snacks and drinks that keep our performers energized.

The grand finale formula:

$$E = I\,Vt$$

Where:

E is Energy (the stamina in joules or kilowatt-hours)
I is Current (our performers again)
V is Voltage (the ever-present ringmaster)
t is Time (how long the show lasts)
Encore!

Our firefly (100W bulb) glowing for 5 hours uses 0.5kWh of energy – a light snack.
The bonfire (2000W heater) blazing for 2 hours devours 4kWh – a feast!
Power and Energy: A Dynamic Duo

Power is how much energy is used right now, while energy is the total used over time. They're like two sides of the same circus coin!

3. Real-World Wonders: The Main Attraction

Smart Homes: Imagine a circus where the lights dim automatically when the performers take a break, and the temperature adjusts to keep everyone comfortable. That's a smart home, using energy wisely!

Factories: In the grand industrial circus, efficient motors and lights ensure no energy is wasted, and machines run like clockwork.
Renewable Energy: The sun and wind join the circus, providing clean energy to keep the show going and going!

4. Advanced Acrobatics: For the Daring!

Power Factor: Sometimes, our performers get out of sync, wasting energy. Power factor correction is like a choreographer, getting them back in rhythm.
Reactive Power: This is the energy used to set up the stage and take it down – essential, but not part of the main performance.
Harmonics: Like a mischievous clown disrupting the show, harmonics distort the smooth flow of energy. Filters are like bouncers, keeping things orderly.
Energy Storage: Like a talented juggler, energy storage keeps energy flowing smoothly, even when the wind isn't blowing or the sun isn't shining.

5. The Future of the Circuit Circus

Electric cars, smart grids, and amazing new batteries are all part of the future of our electrifying circus. Get ready for a show that's even more spectacular, efficient, and sustainable!

Finale:

Understanding power and energy is like being in the audience of the most amazing electrical circus. By appreciating how these concepts work, we can make smarter choices about how we use electricity, ensuring a bright and sustainable future for everyone.

The Kilowatt-Hour: A Shockingly Simple Guide

Imagine electricity as a river flowing through your home. The kilowatt-hour (kWh) is like a bucket you use to measure how much of that river you've actually used to power your stuff. Think of it as the "electricity meter" in the language of your appliances!

Why Should I Care About kWh?

Well, my friend, those kWh are the little guys that determine how much you pay for your electricity bill each month. The more you use, the higher the bill. It's like those arcade games – the more you play, the more tokens you need!

Breaking it Down: The kWh Formula

Don't worry, it's not rocket science! Here's the simple formula:

Energy (kWh) = Power (kW) x Time (hours)
Think of "power" as the strength of the electrical flow (like a powerful showerhead) and "time" as how long you use it.

Real-Life Examples:

The Case of the Light Bulb: Let's say you have a 100-watt light bulb (that's 0.1 kW) and you leave it on for 5 hours. You've used 0.5 kWh of electricity. That's like using half a bucket of our electric river!
The Hungry Heater: Now, imagine a 2-kW heater running for 3 hours. That's a whopping 6 kWh – a much bigger gulp from the river!
kWh in Action:

Kilowatt-hours aren't just about bills. They're used for all sorts of things:

Energy Efficiency: Figuring out which appliances are energy hogs and which one's sip electricity politely.
Renewable Energy: Measuring how much power those solar panels on your roof are generating.
Electric Cars: Knowing how much juice your electric car needs to hit the road.

Tips to Tame Your Electricity Bill:

Switch to Energy-Efficient Appliances: Think of it as upgrading to a low-flow showerhead for your electricity.
Turn Off Lights and Appliances: Don't leave that electric river running when you're not using it!

Embrace Natural Light: Let the sun do its thing and give your lights a break.
Unplug Chargers: Those sneaky chargers keep sipping even when your phones full.

The Bottom Line:

Kilowatt-hours are the key to understanding your electricity usage and saving money on your bill. By being mindful of how much you use, you can be an energy-saving superhero and keep more cash in your pocket!

Electrical safety

The Shocking Truth About Your Home's Electrical System (And How Not to Get Zapped!)
We flip a switch, and poof – light! We plug in our gadgets, and they spring to life. But behind this everyday magic lurks a powerful force that demands respect: electricity. Think of your home's wiring as a network of invisible rivers, flowing with energy. When these rivers stay within their banks (your wires and insulation), all is well. But when things go wrong, you could be in for a nasty shock – literally!

Let's dive into the dangers lurking in your walls and appliances, and learn how to keep your home safe and spark-free.

(a) Damaged Insulation: The Electrical System's Skin

Imagine your body's nervous system without skin – a recipe for disaster, right? Well, electrical wires have their own "skin" called insulation. This protective layer keeps the electrical current contained, flowing where it should. But just like a cut can expose your nerves, damaged insulation exposes live wires, turning everyday objects into potential shock hazards.

Danger Zones:

Electric Shock: A jolt from a live wire can range from a mild tingle to a life-threatening shock. The severity depends on the current's strength

and the path it takes through your body. Think of it like water – a small leak might be annoying, but a burst pipe can cause serious damage.
Fire Hazard: Damaged insulation can lead to short circuits, where the electrical current takes a shortcut, generating intense heat. This can ignite nearby materials faster than you can say "fire extinguisher"!
Fried Gadgets: Exposed wires can also zap your precious electronics, leading to costly repairs or replacements.

Everyday Culprits:

Frayed Cords: That old phone charger with the cracked cable? A prime suspect for exposed wires.
Chewed Wires: Fluffy might look innocent, but those puppy teeth can wreak havoc on electrical cords.
Punctured Cables: DIY gone wrong? Accidentally drilling into a wall can unleash an electrical nightmare.
Time Takes its Toll: Just like we age, insulation can become brittle and crack over time.

A Cautionary Tale:

Picture this: London, 2021. A cozy evening turns to chaos when a frayed extension cord powering a heater sparks a fire, engulfing a living room in flames. The culprit? A cord repeatedly crushed under furniture; its insulation worn away like an old tire.

Stay Safe:

Play Detective: Regularly inspect cords and cables for signs of wear and tear – fraying, cracks, or discoloration.
Handle with Care: Don't bend, crush, or pile heavy objects on electrical cords. Treat them like the lifelines they are.
Strain Relief: Use those little rubber guards on plugs to prevent sharp bends that can damage the cord.
Replace, Replace, replace: When in doubt, throw it out! Any damaged cord should be replaced immediately.
Call the Pros: Leave electrical work to the experts. It's not worth risking your safety (or your home) to save a few bucks.

(b) Overheating Cables: When Things Get Too Hot to Handle

Electrical cables are like the highways of your electrical system, designed to carry a specific amount of current. But when they're overloaded, it's like a traffic jam on a hot summer day – things start to heat up, and not in a good way.

Danger Zones:

Fire Hazard: Overheating cables can melt insulation and ignite surrounding materials, turning your home into an inferno.
Insulation Breakdown: Prolonged overheating can degrade insulation over time, increasing the risk of shocks and short circuits.
Gadget Meltdown: Overheating can also damage your appliances, shortening their lifespan or causing them to malfunction.
Toxic Fumes: Burning insulation releases harmful toxins, adding another layer of danger to an already dangerous situation.

Overheating Hotspots:

Overloaded Circuits: Plugging too many appliances into a single circuit is like trying to cram a whole orchestra into a phone booth – something's got to give.
Loose Connections: Poorly made connections can create electrical bottlenecks, leading to overheating.
Coiled Cords: Coiling an extension cord while in use traps heat, like wrapping your laptop in a blanket.
Damaged Cables: Damaged insulation or internal wires can increase resistance, leading to overheating.

A Cautionary Tale:

Imagine a bustling New York office building in 2023. Suddenly, smoke billows from the server room. The culprit? Overloaded and poorly ventilated cables powering the servers, generating enough heat to ignite nearby insulation and bring business to a screeching halt.

Stay Cool:

Don't Overload: Spread the love! Avoid plugging too many appliances into a single circuit or extension cord.
Right Cable for the Job: Use the correct cable size for the current it needs to carry. Think of it like choosing the right size pipe for your water flow.
Tight Connections: Make sure all connections are secure and properly made.
Ventilation is Key: Provide adequate ventilation around electrical equipment and cables to prevent heat buildup.
Regular Checkups: Keep an eye out for signs of overheating, like discoloration, melting, or a burning smell.

(c) Damp Conditions: Electricity and Water Don't Mix

We all know water and electricity don't mix, but it's easy to forget how often they come into contact in our homes. Think leaky faucets, spilled drinks, and humid bathrooms.

Danger Zones:

Electric Shock: Contact with electrical equipment or water that's become energized due to a fault can deliver a potentially fatal shock.
Equipment Damage: Moisture can corrode wiring and damage electrical components, increasing the risk of hazards.

Water Hazard Zones:

Bathrooms: With water everywhere and wet hands and feet, bathrooms are prime locations for electrical accidents.
Kitchens: Spills, splashes, and appliances near sinks make kitchens another high-risk area.
Outdoor Electrical: Exposed outlets and fixtures are vulnerable to rain and moisture.
Flooding: Floodwaters can turn your entire electrical system into a deadly trap.

A Cautionary Tale:

A Florida construction worker in 2022 paid the ultimate price for working in a damp basement with a faulty power tool. Contact with a wet surface turned a routine task into a tragedy.

Stay Dry:

GFCI Protection: Install Ground Fault Circuit Interrupters (GFCIs) in bathrooms, kitchens, and other wet areas. These life-saving devices quickly shut off power in the event of a ground fault.
Weatherproof Outdoor Electrics: Use weatherproof covers and installations for outdoor outlets and switches.
Appliance Safety: Keep electrical appliances away from water sources and ensure they're properly grounded.
Dry Hands: Always dry your hands before handling electrical equipment or plugs.
Avoid Water: Never use electrical appliances while standing in water or on a wet surface.

(d) Excess Current from Overloading: Too Much of a Good Thing

Electricity is like food for our appliances – they need it to function. But just like overeating can strain your digestive system, overloading your electrical circuits can lead to serious problems.

Danger Zones:

Overheating: Overloading causes cables, plugs, sockets, and appliances to overheat, increasing the risk of fire and damage.
Fire Hazard: Overheating from overloading is a common cause of electrical fires.
Equipment Damage: Excessive current can damage appliances and shorten their lifespan.
Tripped Breakers: Overloading can trip circuit breakers, interrupting power to protect the circuit from damage.

Overloading Traps:

Appliance Overload: Plugging multiple high-powered appliances (like heaters, kettles, and hair dryers) into a single outlet or extension cord is a recipe for disaster.
Daisy-Chaining: Connecting multiple extension cords together reduces the current-carrying capacity of each cord, increasing the risk of overloading.
Faulty Appliances: Appliances with faulty wiring or components can draw excessive current, overloading the circuit.
Old Wiring: Older electrical wiring may not be able to handle the demands of modern appliances.

A Cautionary Tale:

In a California student dormitory in 2024, a power strip overloaded with laptops, phone chargers, and other devices sparked a fire, injuring students and causing significant damage.

Stay Within Limits:

Distribute the Load: Avoid plugging too many appliances into a single outlet or extension cord.
Right Circuit for the Job: Ensure high-powered appliances are plugged into circuits designed to handle their load.
No Daisy-Chaining: Avoid connecting multiple extension cords together.
Surge Protection: Surge protectors can help protect appliances from damage caused by power surges and overloading.
Appliance Checkups: Regularly check appliances for signs of damage or wear that could lead to overloading.

The Bottom Line:

Electricity powers our modern lives, but it's a force to be reckoned with. By understanding the dangers of damaged insulation, overheating cables, damp conditions, and overloading, you can take steps to protect yourself, your loved ones, and your home.

⚡ Zapping Dangers and the Tiny Heroes That Save Us! ⚡

Ever felt that little jolt when you touch a faulty appliance? That's your body getting a taste of the raw power coursing through our homes! Luckily, we have tiny but mighty heroes standing guard: trip switches and fuses! These unsung champions of safety work tirelessly to prevent electrical mayhem. Let's dive into their world and uncover their secrets!

1. Meet the Guardians: Trip Switches & Fuses

Think of your electrical system as a network of roads with electricity as the traffic. Sometimes, there's a surge of "traffic" (overload) or a sudden "crash" (short circuit). That's where our heroes come in!

1.1 Trip Switches: The Resettable Protectors

These clever devices are like traffic lights that turn red when there's trouble. They sense danger and instantly cut off the power, preventing damage to your appliances and keeping you safe from shocks. The best part? They're reusable! Just flip the switch back on, and you're good to go.

Miniature Circuit Breakers (MCBs): These are the everyday heroes found in most homes, protecting against overloads and short circuits.
Residual Current Devices (RCDs): Imagine a super-sensitive detective that can detect even the tiniest leak in the electrical "pipes." That's an RCD! They protect you from dangerous electric shocks caused by faulty appliances or wiring.
RCBOs: The ultimate multitasker! These combine the powers of MCBs and RCDs in one compact device.

1.2 Fuses: The Self-Sacrificing Warriors

Fuses are like brave knights that sacrifice themselves to protect the kingdom. When danger strikes, they bravely melt, breaking the electrical circuit and stopping the flow of power. They're single-use heroes, but their sacrifice prevents fires and keeps you safe.

Choosing the Right Hero:

Just like superheroes have different powers, trip switches and fuses come in different strengths. The right choice depends on the "traffic" (electrical load) on each circuit. Think of it like choosing the right size pipe for water flow.

1.3 Trip Switch vs. Fuse: A Friendly Duel

Both are awesome, but they have different strengths:

Feature	Trip Switch	Fuse
Reusability	Reusable (yay!)	Single-use (sniff...)
Resetting	Easy	Needs replacement
Speed	Lightning-fast	A bit slower
Sensitivity	Super sensitive	Less sensitive
Cost	A bit pricier	Budget-friendly

2. When Electricity Goes Rogue: Live Wire Meets Earthed Metal

Imagine a live wire, buzzing with electricity, accidentally touching a metal casing. Yikes! This is an earth fault, and it's a recipe for disaster.

What Happens:

Electrical Flood: A huge surge of electricity flows through the metal casing towards the earth.
Heroic Intervention: Trip switches or fuses sense the danger and instantly cut off the power.
Safety First: The faulty circuit is isolated, preventing further danger.

Earthing: The Safety Net

Earthing is like a safety net for your electrical system. It provides a safe path for electricity to flow to the earth in case of a fault, ensuring that the protective devices work quickly and effectively.

The Dangers:

Without proper protection, earth faults can lead to:

Electric Shocks: Touching the faulty appliance can give you a nasty jolt, potentially causing serious injury.
Fire Hazard: The high current flow can generate intense heat, leading to fires.
Equipment Damage: Your precious appliances can be fried in the process.

A Real-Life Story:

Imagine a faulty washing machine with a damaged cord. The live wire touches the metal casing, and suddenly, there's a surge of electricity! Thankfully, the RCD springs into action, cutting off the power before anyone gets hurt. Phew!

The Grand Finale: Stay Safe, My Friends!

Trip switches and fuses are the silent guardians of our electrical world. They protect us from shocks, fires, and equipment damage. By understanding how they work and taking necessary precautions, we can all enjoy the benefits of electricity without the fear of zapping dangers!

Imagine Electricity as a Wild River:

Think of electricity like a powerful, untamed river. It's a source of immense energy and potential, but if not properly controlled, it can be incredibly dangerous. Our homes and appliances are like the structures built along this riverbank, and electrical safety measures are the dams, levees, and channels that keep the river from overflowing and causing destruction.

1. The Appliance Casing: Your Personal Life Raft:

The casing of an electrical appliance is like your personal life raft on this electric river. If the appliance malfunctions and the "river" (electricity) spills out, you want to be sure your raft is secure and won't let you get swept away.

Double Insulation: A Raft with Extra Buoyancy: This is like having a super strong, double-layered raft. Even if one layer gets punctured, the other keeps you afloat and safe from the electric current. Think of it as the difference between a flimsy inflatable pool toy and a sturdy kayak.

Earthing: A Safety Line to Shore: Imagine your raft has a strong rope tethered to the shore. If the electrical current surges, this rope acts as a grounding wire, pulling the excess current safely into the earth, like a lightning rod diverting a bolt of lightning.

2. The Switch: The Dam That Controls the Flow:

The switch is like a dam on our electric river. It gives you the power to control the flow of electricity.

Live Wire Connection: Shutting off the Source: Imagine the switch is built directly at the source of the river. By connecting it to the live wire, you're effectively shutting off the water flow at its origin, ensuring nothing can get past even if there's a leak downstream.

3. Fuses and Circuit Breakers: The Emergency Spillways:

Fuses and circuit breakers are like emergency spillways on a dam. If the current becomes too strong, like a flood threatening to overwhelm the dam, these safety devices "open the spillway" and redirect the excess current, preventing damage to the system.
Live Wire Connection: Containing the Flood: By connecting these safety devices to the live wire, you ensure that if the "floodgates" open, the flow is cut off right at the source, preventing any further damage.

In Conclusion:
Electrical safety is all about respecting the power of electricity and taking the necessary precautions to harness it safely. By understanding the "why" behind these measures, we can better appreciate their importance and make informed decisions to protect ourselves and others.

Electromagnetic effects
Electromagnetic induction

The Invisible Dance of Magnets and Metal: A Tale of Attraction and Opposition

Imagine a world where magnets come alive, their invisible forces weaving a story of push and pull, of energy and resistance. Let's dive into this world and witness the captivating dance of electromagnetic induction and Lenz's Law.

Act 1: The Magnet's Enchantment

Our stage is set with a simple coil of copper wire, a delicate galvanometer, and a powerful bar magnet – the star of our show. As the magnet approaches the coil, it's as if an invisible hand awakens within the wire, compelling a surge of electrons to flow. The galvanometer's needle twitches, a testament to this unseen current.

But like a fleeting enchantment, the current vanishes when the magnet rests. And as the magnet retreats, the needle dances again, but this time in the opposite direction, as if protesting the magnet's departure.

This is the magic of Faraday's Law: the changing magnetic field conjures an electromotive force, a whisper of energy that sets the electrons in motion.

Act 2: Lenz's Law – The Rebellious Spirit

But there's another force at play, a rebellious spirit that opposes the magnet's influence. This is Lenz's Law, the guardian of energy conservation.

Imagine dropping a magnet through a copper tube. It's like watching a feather drift through the air, defying gravity's pull. The magnet descends with an unexpected grace, as if resisting its natural course.

This defiance stems from the eddy currents, swirling whirlpools of electrons within the copper, awakened by the falling magnet. These currents, like tiny defenders, create their own magnetic field, pushing back against the magnet's intrusion.

The Grand Finale: Applications of this Enchanting Dance

This invisible dance of magnets and metal isn't just a spectacle; it's the driving force behind many marvels of our modern world.

Electric Generators: Imagine a tireless dancer, spinning within a magnetic field, generating a continuous flow of energy. This is the essence of an electric generator, converting mechanical motion into electrical power.

Transformers: These silent guardians regulate the flow of electricity, stepping up or down the voltage, ensuring a harmonious balance in our power grids.

Magnetic Brakes: Like an invisible hand gently restraining a speeding train, magnetic braking systems provide a smooth and controlled halt, harnessing the power of Lenz's Law.

Induction Cooking: The culinary stage is set with a mesmerizing dance of magnetic fields and cookware. Eddy currents heat the pot directly, offering a swift and efficient cooking experience.

Beyond the Stage:

The dance of electromagnetic induction and Lenz's Law extends far beyond these examples. It's a fundamental principle woven into the fabric of our technological world, from medical imaging to wireless communication.

By understanding this captivating interplay of forces, we unlock a deeper appreciation for the unseen wonders that shape our lives. It's a reminder that even in the seemingly mundane, there's magic waiting to be discovered.

The Dance of Electrons and Magnets: A Tale of Induced EMF

Imagine a conductor, a wire perhaps, twirling gracefully within the invisible embrace of a magnetic field. As it pirouettes, a curious phenomenon unfolds: an electromotive force, or EMF, is born within the wire, like a spark of life. This is electromagnetic induction, a captivating ballet of electrons and magnets.

But what determines the strength of this induced EMF, this electrical awakening? Let's delve into the heart of this magnetic choreography, where two key factors take center stage.

The Tempo of Change: A Magnetic Tango

Picture the magnetic field as a restless sea, its lines of force ebbing and flowing. The faster these lines shift and swirl, the more they buffet the electrons within the conductor, urging them into a lively dance. This quickening of the magnetic field, this acceleration of change, is the first key to a stronger induced EMF.

Think of a magnet darting towards a coil of wire. The swifter the magnet's approach, the more dramatic the change in the magnetic field, and the greater the surge of EMF within the coil. It's like a crescendo in an orchestra, the music swelling as the tempo increases.

The Chorus of Coils: Strength in Numbers

Now, imagine not just one dancer, but a multitude, each mirroring the other's movements. This is the essence of the second factor: the number of turns in a coil. Each turn acts as a separate stage for the magnetic ballet, generating its own EMF. These individual sparks of energy join forces, amplifying the overall electrical current.

Think of a solenoid, a tightly wound coil resembling a spring. As a magnet glides through its core, each turn of the solenoid contributes to the induced EMF, creating a chorus of electrical energy. It's like adding more voices to a choir, the sound growing richer and more powerful.

Real-World Rhapsodies: From Generators to MRI Machines

These principles of electromagnetic induction aren't confined to the theoretical stage. They power the real world in countless ways.

Consider the electric generator, a workhorse of modern civilization. Within its core, a coil spins tirelessly within a magnetic field. The faster the rotation, the more frenzied the dance of electrons, and the greater the electrical output. It's like a turbine transforming the raw energy of flowing water into a symphony of electricity.

Or take the MRI machine, a marvel of medical technology. Its powerful magnets and intricate coils orchestrate a delicate dance of protons within the human body, generating signals that reveal our inner landscapes. It's like composing a detailed map of the body's hidden realms, guided by the principles of electromagnetic induction.

Beyond the Basics: A World of Wonders

The dance of electrons and magnets is far from simple. Other factors, like the strength of the magnetic field and the orientation of the coil, also influence the induced EMF. It's a complex choreography, with endless variations and nuances.

But by understanding the fundamental steps, the tempo of change and the chorus of coils, we can unlock the secrets of this magnetic ballet. We can harness its power to illuminate our homes, explore the universe, and even peer into the depths of the human body.

So let us continue to marvel at this intricate dance, this interplay of electrons and magnets, and explore the boundless possibilities it holds for our technological future.

The A.C. generator

AC Generators: The Unsung Heroes of Our Electric World (2024 Update)

Imagine a world without that satisfying of electricity powering your devices. No Netflix binges, no late-night coding sessions, no refrigerator humming along. It's a bleak picture, right? Well, thank goodness for AC generators, the tireless workhorses that keep our modern lives humming.

These ingenious machines, born from the mind-bending magic of electromagnetic induction, are like tiny power plants, transforming mechanical energy (think spinning turbines) into the electrical lifeblood that courses through our homes and cities.

Let's embark on a journey to unravel the secrets of these electrifying contraptions, diving deep into their inner workings, their mesmerizing outputs, and the myriad ways they power our world.

1. The Heart of the Machine: How AC Generators Work

At its core, an AC generator is a beautiful dance between a coil of wire and a magnetic field. Picture a wire hula hoop twirling gracefully within the invisible embrace of a magnet. This can be achieved in a couple of ways:

Rotating Coil, Steady Magnet: Imagine our hula hoop (the coil) spinning between the unwavering gaze of two magnets.
Rotating Magnet, Steady Coil: Now, picture the magnets doing a pirouette around the stationary hula hoop. This is often preferred in massive power plants, as it makes it easier to tap into the electrical current.
But regardless of who's spinning, the underlying principle is the same: Faraday's Law of Electromagnetic Induction. This law, a cornerstone of electromagnetism, tells us that a changing magnetic field around a wire creates an electric current, like a magnetic whisper urging electrons to flow. As the coil or magnet spins, this magnetic field constantly shifts, giving birth to an alternating current that ebbs and flows like the tide.

Slip Rings and Brushes: The Dynamic Duo

To harness this electrical current and send it out into the world, we need a clever connection system: slip rings and brushes. Imagine two shiny rings attached to our spinning hula hoop. These slip rings glide against stationary carbon brushes, creating a continuous electrical pathway from the whirling coil to the outside world. It's like a high-five between the generator and the electrical grid, allowing the alternating current to flow freely.

The Dance of Electricity: A Step-by-Step Guide

Let's break down the mesmerizing dance of electricity generation:

Starting Position: Imagine the hula hoop perfectly aligned with the magnetic field. At this moment, the magnetic force is at its peak, but it's not changing, so no current is generated. It's like a dancer poised for action.
The Spin Begins: As the coil rotates, the magnetic field around it starts to shift, like a spotlight changing direction. This change induces an electric current, bringing the dancer to life.
Maximum Power: When the hula hoop is at a right angle to the magnetic field, the change in magnetic force is most dramatic, leading to the peak electrical current. The dancer is in full swing!
Back to Zero: As the coil continues to spin, the current gradually diminishes, returning to zero when the coil is once again aligned with the magnetic field. The dancer takes a bow.
Reverse the Flow: The coil keeps spinning, but now the current flows in the opposite direction, reaching a peak when the coil is again perpendicular to the magnetic field, but facing the other way. The dancer has changed direction!
The Cycle Repeats: The coil returns to its starting position, and the entire process begins anew, creating a continuous flow of alternating current. The dance goes on!

2. Visualizing the Electric Symphony: The Output Waveform

The electrical output of an AC generator can be beautifully captured in a graph, where the current dances across time in a graceful sine wave.

This wave, with its smooth peaks and valleys, tells the story of the alternating current.

Peaks and Troughs: The peaks of the wave represent the maximum current, while the troughs show the minimum (which is just the peak in the opposite direction). It's like the dancer's highest leap and lowest dip.
Zero Crossings: The points where the wave crosses the zero line represent moments of no current, like the dancer pausing between steps.
This mesmerizing sine wave reveals the dynamic nature of alternating current, constantly changing in magnitude and direction. The speed of the coil's rotation dictates the frequency of the wave, or how quickly the dancer completes a cycle.

3. Beyond the Basics: Advanced Concepts and Real-World Wonders

The strength of the electrical current generated depends on several factors:

Magnetic Muscle: A stronger magnetic field leads to a more powerful current, like a more energetic dancer.
Coil Size: A larger coil captures more of the magnetic field, resulting in a greater current. Think of a bigger hula hoop.
Number of Turns: More turns in the coil mean more interaction with the magnetic field, leading to a stronger current. It's like adding more dancers to the performance.
Speed of Rotation: Faster rotation means a more rapid change in the magnetic field, and thus a stronger current and higher frequency. The dancer is really twirling now!

Types of AC Generators:

Synchronous Generators: These generators produce AC power at a fixed frequency, perfectly synchronized with the rotation speed. They are the backbone of power plants, like a conductor leading an orchestra.
Induction Generators: These generators operate without needing to be perfectly synchronized, making them ideal for wind turbines and other

renewable energy sources. It's like a freestyle dance within the overall performance.

AC Generators in Action: Powering Our World

AC generators are the unsung heroes of our modern world, quietly working behind the scenes to power our lives:

Power Plants: Giant AC generators in power plants transform various energy sources, like fossil fuels, nuclear power, and the raw power of nature, into the electricity that flows to our homes.
Transportation: AC generators, also known as alternators, keep our car batteries charged and power our headlights and radios.
Portable Power: Smaller generators provide electricity during emergencies or in remote locations, like a lifeline in times of need.
Renewable Energy: AC generators are essential for harnessing the power of wind, water, and sunlight, bringing us closer to a sustainable future.

Looking Ahead: The Future of AC Generators

As technology continues to evolve, so too will AC generators:

Smart Grid Integration: Imagine generators that can seamlessly communicate with the power grid, optimizing energy flow and preventing blackouts.
Advanced Materials: New materials, like superconductors, could make generators even more efficient, like giving our dancer super-powered shoes.
Miniaturization: Tiny AC generators could power miniature devices and sensors, opening up exciting new possibilities.

In Conclusion:

AC generators are the silent giants of our electric world, tirelessly converting mechanical energy into the electrical power that drives our lives. From the massive turbines in power plants to the alternator in your car, these ingenious machines are essential to our modern existence. As we move towards a future powered by clean and sustainable energy, AC generators will continue to play a vital role, ensuring that the lights stay on and the world keeps spinning.

Magnetic effect of a current

The Magnetic Tango of Electrons: A Whimsical Journey into the Heart of Electromagnetism

Imagine tiny dancers, twirling and swirling in an invisible ballroom. These dancers are electrons, and their graceful movements create a mesmerizing spectacle – the magnetic field. This enchanting phenomenon, where electric currents give birth to magnetic fields, is the cornerstone of electromagnetism, the magic behind everything from the humble fridge magnet to the mind-boggling MRI machine.

1. The Electric Current's Choreography: Straight Wires and Solenoids

Straight Wires: A Solo Performance

Picture a single electron, gliding along a straight wire like a ballerina on a tightrope. As it moves, it creates a whirlwind of magnetic energy around it, like a swirling skirt. This magnetic field forms invisible rings around the wire, with the wire itself as the central axis.

To understand the direction of this magnetic swirl, we use the "Right-Hand Rule," a secret handshake of physics:

The Right-Hand Rule: Pretend you're holding the wire with your right hand, thumb pointing in the direction of the electron flow. The way your fingers curl around the wire reveals the direction of the magnetic field.
Key characteristics of this magnetic performance:

Circular Waltz: The magnetic field lines form closed loops, like dancers in a never-ending waltz around the wire.
Fading Applause: The further you move from the wire, the weaker the magnetic field becomes. It's like the applause fading as you leave the theater.

Perpendicular Pirouette: The magnetic field always pirouettes at a right angle to the direction of the current.

Factors influencing the magnetic performance:

The Tempo of the Current: The strength of the magnetic field depends on how fast the electrons are moving (the current). A faster tempo means a stronger magnetic field.
The Direction of the Dance: Reversing the direction of the current is like reversing the dance; it reverses the direction of the magnetic field.

Solenoids: A Chorus Line

Now, imagine many wires, not just one, all coiled together like a chorus line. This is a solenoid. When an electric current flows through this coil, it creates a magnetic field that mimics a bar magnet. Each loop in the coil adds its own magnetic voice to the chorus, creating a powerful and harmonious magnetic field within the solenoid.

Key characteristics of this magnetic chorus:

Uniform Harmony: Inside the solenoid, the magnetic field is uniform, like a well-rehearsed chorus line, with the same strength and direction throughout.
Magnetic Embrace: The magnetic field lines inside the solenoid are parallel and tightly packed, like dancers in a close embrace. Outside the solenoid, they spread out like a blooming flower.
North and South Poles: The solenoid, like a bar magnet, has a north pole at one end and a south pole at the other. The magnetic field lines flow from north to south outside the solenoid and from south to north inside.

Factors influencing the magnetic chorus:

The Current's Tempo: As with the single wire, increasing the current strengthens the magnetic field.
The Number of Dancers: The more turns in the coil (more dancers in the chorus line), the stronger the magnetic field.

The Stage: Introducing a ferromagnetic core, like iron, into the solenoid is like giving the dancers a special stage. It amplifies the magnetic field significantly.

Visualizing the Invisible Dance:

To see this invisible dance of the magnetic field, we can use iron filings. When sprinkled around a current-carrying conductor, they align themselves along the magnetic field lines, like tiny compasses revealing the secret choreography of the magnetic field.

2. The Magnetic Stage: Relays and Loudspeakers

The magnetic effect of a current isn't just a theoretical dance; it powers many devices we use every day. Let's take a look at two-star performers on the magnetic stage: relays and loudspeakers.

Relays: The Electric Switchboard Operator

A relay is like a switchboard operator, using the magnetic field to control electric circuits. It's an electrically operated switch with an electromagnet, an armature, and a set of electrical contacts. When current flows through the electromagnet's coil, it creates a magnetic field that pulls the armature, like a dancer responding to a cue. This movement opens or closes the electrical contacts, directing the flow of electricity in another circuit.

Key components of the relay:

Electromagnet: The coil of wire that creates the magnetic field, the dancer's cue.
Armature: The movable piece of iron or steel that responds to the magnetic field, the dancer.
Contacts: The electrical terminals that are opened or closed by the armature, the switchboard.

How the relay works:

Cue the Electromagnet: When current flows through the electromagnet's coil, it creates a magnetic field.

The Dancer Responds: The magnetic field attracts the armature, making it move.
Switching the Connections: The armature's movement opens or closes the electrical contacts.
Controlling the Performance: This controls the flow of current in another circuit, perhaps a high-power one.

Advantages of using relays:

Isolation: Relays keep the control circuit and the controlled circuit separate, preventing any electrical mishaps.
Amplification: They can control a high-power circuit with a low-power signal, like a conductor leading a powerful orchestra.
Multitasking: A single relay can control multiple circuits simultaneously, a true multitasker.

Applications of relays:

Cars: Controlling headlights, starter motors, and windshield wipers.
Factories: Controlling motors, pumps, and other industrial equipment.
Telephones: Historically used in telephone exchanges for routing calls.
Household Appliances: Found in refrigerators, washing machines, and air conditioners.

Example: The Car Starter Relay

In a car starter system, the relay is the key to starting the engine. When you turn the ignition key, a small current flow through the relay's coil, energizing the electromagnet. The electromagnet attracts the armature, closing the contacts and allowing a large current to flow from the battery to the starter motor, bringing the engine to life.

Loudspeakers: The Sound Magicians

A loudspeaker is a magician, transforming electrical signals into the sounds we hear. It uses the magnetic effect of a current to move a diaphragm, which vibrates the air, creating sound waves.

Key components of the loudspeaker:

Voice Coil: The coil of wire attached to the diaphragm, the magician's wand.
Permanent Magnet: The strong magnet that provides a static magnetic field, the magician's stage.
Diaphragm: The cone-shaped structure that vibrates to produce sound waves, the magician's voice.

How the loudspeaker works:

The Electrical Spell: An electrical audio signal is sent to the voice coil.
Magnetic Interaction: The current in the voice coil creates a magnetic field that interacts with the permanent magnet's field.
The Dance of the Voice Coil: This interaction creates a force, making the voice coil move back and forth.
Vibrating the Air: The voice coil's movement vibrates the diaphragm.
The Sound Emerges: The diaphragm's vibration creates pressure waves in the air, which we perceive as sound.

Factors influencing the loudspeaker's magic:

The Magnet's Power: A stronger magnet creates a stronger force on the voice coil, leading to louder sound.
The Diaphragm's Shape and Size: The diaphragm's size and shape affect the frequency response. Larger diaphragms are better at low frequencies, while smaller ones excel at high frequencies.
The Diaphragm's Material: The material affects the sound quality. Different materials have different stiffness and damping, influencing the diaphragm's vibrations.

Applications of loudspeakers:

Home Audio: Stereo systems, home theater systems.
Public Address Systems: Stadiums, concert halls, auditoriums.
Telephones: Converting electrical signals into audible speech.
Mobile Devices: Smartphones, tablets, laptops.
Example: The Smartphone Loudspeaker

The loudspeaker in your smartphone is a miniature version of a traditional loudspeaker. It uses a tiny voice coil, a small permanent magnet, and a thin diaphragm to produce sound. The electrical audio signal from the phone's amplifier is sent to the voice coil, which interacts with the magnetic field of the permanent magnet to move the diaphragm, creating the sound you hear.

Conclusion: The Magnetic Symphony

The magnetic effect of a current is a fundamental principle of electromagnetism, a beautiful dance of electrons that powers countless devices in our daily lives. From simple electromagnets to complex devices like relays and loudspeakers, this phenomenon plays a crucial role in our world. Understanding the principles behind this magnetic symphony allows us to appreciate the ingenuity of these inventions and paves the way for future innovations in electrical and electronic technologies.

Forces on a current-carrying conductor

Imagine a Dance of Invisible Forces

Have you ever seen a ballet dancer leap and twirl, seemingly defying gravity? That's the magic of physics in motion, just like the unseen forces that make electricity dance. Now, picture this: a wire, humming with electrical energy, suddenly comes alive in the presence of a magnet. It's as if an invisible hand gives it a nudge, making it move. This, my friend, is the captivating world of electromagnetism!

A Playful Experiment: The Wire That Dances

Let's step into a science lab and witness this magic firsthand. We'll need a few simple props: a powerful horseshoe magnet, a copper wire like a tightrope, a source of electrical current, and some clips to connect everything. Imagine the copper wire as our dancer, suspended between the arms of the magnet. As we switch on the current, it's like turning on the music. Our wire starts to move, swaying to the rhythm of the electromagnetic force!

The Secret Handshake: Fleming's Left-Hand Rule

But how does this dance work? It's all about the secret handshake between electricity and magnetism. Imagine your left hand as the choreographer of this dance. Your thumb points in the direction the wire moves, your index finger aligns with the magnetic field, and your middle finger shows the direction of the electric current. This is Fleming's Left-Hand Rule, a handy trick to predict how the wire will move.

Magnetic Tango: When Wires Dance Together

Now, let's introduce another wire to our stage. When two wires carry electricity, they create their own magnetic fields, like partners in a tango. If the currents flow in the same direction, they attract, pulling closer as if whispering secrets. But if the currents flow in opposite directions, they repel, pushing away like rivals in a dramatic standoff.

Electromagnetism in Action: From Motors to Maglev Trains

This playful dance of electromagnetism isn't just a laboratory curiosity. It's the driving force behind countless technologies we use every day. Electric motors, the heart of many machines, use this force to spin and create motion. Generators do the opposite, turning motion into electricity, like a magical windmill. Loudspeakers use this dance to create sound waves, transforming electrical signals into music that moves our souls. And maglev trains? They levitate above the tracks, gliding effortlessly on a cushion of magnetic forces, taking us into the future of transportation.

The Railgun: A Force to be Reckoned With

Imagine a weapon that can launch projectiles at incredible speeds, like a futuristic slingshot. That's the power of the railgun, a marvel of electromagnetism. It uses parallel rails and a surge of electricity to create a powerful magnetic field that propels a projectile forward. This technology has potential applications in defense, space exploration, and scientific research, pushing the boundaries of what's possible.

The Grand Finale: A World of Wonder

The force that makes our wire dance is not just a scientific concept; it's a testament to the interconnectedness of the universe. It's a reminder that the world is full of hidden forces, waiting to be discovered and harnessed for the betterment of humanity. So, the next time you see an electric motor whirring or hear music blasting from a speaker, remember the invisible dance of electromagnetism, a force that shapes our world in countless ways.

The D.C. motors

Imagine a Dance of Invisible Forces: The Magic of Electric Motors

Have you ever wondered how a tiny battery can make a toy car zoom across the floor? Or how a washing machine spins with such relentless energy? The answer lies in a captivating dance between electricity and magnetism, a phenomenon that powers our world in countless ways.

The Secret Language of Magnets and Wires

Imagine a wire, buzzing with tiny electrons, like a busy highway. Now, picture a magnet nearby, with its invisible magnetic field, like an aura surrounding it. When the wire with its flowing current enters this magnetic field, something amazing happens! It's as if the wire feels a gentle nudge, a force pushing it to move. This is the "motor effect," a fundamental principle of physics that makes electric motors possible.

A Twist and a Shout: The Birth of Torque

Now, let's coil that wire, like a spring, and place it within the magnet's embrace. The forces on the wire don't just push; they create a twisting, turning force called torque. It's like the wire wants to dance, to spin and twirl within the magnetic field. The more turns in the coil, the stronger the current, and the more powerful the magnet, the more enthusiastic this dance becomes!

The Commutator: A Clever Choreographer

But how do we keep this dance going, turning it into a continuous spin? Enter the commutator, a clever device that acts like a choreographer. It constantly switches the direction of the current in the coil, ensuring that the twisting force never stops. It's like changing partners in a dance, keeping the energy alive and the motion flowing.

The Electric Motor: A Symphony of Motion

And there you have it – the electric motor, a marvel of engineering that converts electrical energy into mechanical motion. It's like a tiny orchestra, with the magnet as the conductor, the coil as the dancer, and the commutator as the choreographer, all working together to create a symphony of motion.

From Tiny Toys to Mighty Machines

Electric motors are everywhere, from the tiny vibrations in your phone to the powerful engines in electric cars. They power our homes, our industries, and our transportation systems. They are the silent workhorses of modern society, driving progress and making our lives easier and more enjoyable.

A World of Possibilities

As we continue to explore the fascinating interplay of electricity and magnetism, we can imagine even more incredible applications for electric motors. From robots that perform delicate surgeries to flying cars that soar through the skies, the possibilities are endless. So next time you encounter an electric motor, take a moment to appreciate the intricate dance of forces that makes it all possible. It's a testament to human ingenuity and our endless quest to harness the power of nature.

The transformer

The Unsung Hero of Electricity: A Tale of the Iron-Cored Transformer

Imagine a bustling city, lights ablaze, machines humming. None of it would be possible without the humble transformer, quietly working behind the scenes. This isn't some caped crusader, but a chunk of iron and wire that performs a kind of magic: shifting electricity from high voltage to low, or vice versa.

Think of it like a gear system for electricity. A step-up transformer is like shifting into high gear for long distances – it boosts the voltage to send power efficiently across miles and miles of power lines. Then, a step-down transformer shifts down to deliver a safe, usable voltage to your home.

Inside the Magic Box:

Let's peek under the hood. We've got two coils of wire, like intertwined slinkies, wrapped around a core of iron. When electricity flows through the first coil (the primary winding), it creates an invisible force field – a magnetic field. This field flows through the iron core, like water through a pipe, and causes a current to appear in the second coil (the secondary winding). Voilà! Voltage transformed.

Why Iron? Why Not, Say, a Carrot?

Iron is a magnet's best friend. It concentrates the magnetic field, making the energy transfer super-efficient. A carrot, sadly, wouldn't be much help.

The Transformer's Superpowers:

Efficiency: These things are incredibly efficient, losing very little energy in the process. It's like getting 95 miles to the gallon!
Shapeshifting Electricity: Need high voltage for transmission? No problem. Need low voltage for your phone charger? Easy.
Guardian Angel: Transformers can isolate circuits, protecting your precious electronics from power surges.

Transformers in Action:

Power Grid: They're the backbone of the electricity grid, making sure power gets where it needs to go.
Music to Your Ears: Audio transformers help your music sound its best by matching the amplifier to your speakers.
Renewable Energy: They're key players in solar and wind power, getting that green energy onto the grid.

The Future is Electric (and Transformative):

Scientists are always tinkering, making transformers even better. Imagine transformers that can talk to the grid, adjusting themselves to keep things running smoothly, or super-efficient transformers made with newfangled materials.

So next time you flip a light switch, remember the unsung hero – the iron-cored transformer – silently shaping the flow of electricity that powers our modern world.

Transformers: The Unsung Heroes of Electricity Delivery

Imagine electricity as a bustling city, with power plants as the central hubs and homes and businesses as the suburbs. Just like we need efficient highways to connect the city center to the outskirts, we need a reliable system to transport electricity over long distances. That's where transformers come in – they're the "gear shifters" of the electrical grid, ensuring power flows smoothly and efficiently.

The Magic of Transformers: A Tale of Two Coils

At the heart of a transformer lies a simple yet ingenious concept: two coils of wire wrapped around a magnetic core. When electricity flows through the first coil (the primary coil), it creates a magnetic field that dances across to the second coil (the secondary coil). This magnetic tango induces an electric current in the secondary coil, but here's the kicker – the voltage can be magically stepped up or down depending on the number of turns in each coil.

Think of it like a seesaw: more turns on the secondary side, and the voltage goes up (step-up transformer); fewer turns, and the voltage goes down (step-down transformer). It's like having a volume knob for electricity!

Why High Voltage is the Holy Grail of Power Transmission

Now, imagine trying to send electricity through hundreds of miles of wires. It's like trying to squeeze a massive crowd through a narrow doorway – things are bound to get messy and inefficient. That's where high-voltage transmission comes in.

By cranking up the voltage, we can dramatically reduce the amount of current needed to transmit the same amount of power. Why is this a big deal? Because electrical resistance, the "friction" in the wires, loves to gobble up power and spit it out as wasted heat. Lower current means less resistance, and less resistance means less energy lost along the way.

The "Ohm-y" Truth About Power Loss

Remember Ohm's Law? It's like the traffic rule of electricity, stating that voltage (V) equals current (I) times resistance (R). Power loss in transmission lines is proportional to the square of the current. So, if you halve the current, you slash the power loss by a whopping 75%!

High Voltage: Saving the Planet, One Electron at a Time

High-voltage transmission isn't just about efficiency; it's also about sustainability. By minimizing energy waste, we reduce our reliance on fossil fuels and lessen our environmental impact. Plus, high-voltage lines can carry more power, allowing us to tap into renewable energy sources like wind and solar farms located far away from cities.

Real-World Transformers: From the UK to China

Transformers and high-voltage transmission are the backbone of modern power grids worldwide. The UK's National Grid, for example, uses a network of high-voltage lines to shuttle power from remote power stations to bustling cities. China is pushing the boundaries with

Ultra-High Voltage (UHV) lines, exceeding 800kV, to transport clean energy across vast distances.

Transformers: The Silent Guardians of Our Electrified World

So, the next time you flip a switch, remember the unsung heroes working behind the scenes – the transformers diligently humming in substations, ensuring that electricity reaches you safely and efficiently. They're the silent guardians of our electrified world, making modern life possible.

Uses of an oscilloscope

Unlocking the Secrets of Electrical Signals: Your Oscilloscope Adventure

Imagine a device that lets you see electricity – not as a static spark, but as a dynamic, dancing wave. That's the magic of an oscilloscope! It's like a detective's magnifying glass for electronic signals, revealing their hidden shapes and whispers.

1. Painting Waves of Light

Think of an oscilloscope as an artist, with a screen as its canvas and an electron beam as its brush. This beam dances across the screen, guided by the electrical signal it's trying to capture.

Vertical Dance: The signal's voltage pulls the beam up and down, like a puppeteer controlling a marionette. Stronger voltage, higher the dance!
Horizontal Sweep: Meanwhile, a built-in timer pulls the beam from left to right at a steady pace. It's like the canvas itself is moving, creating a timeline for the dance.
The result? A glowing line on the screen that traces the signal's journey through time. It's like a graph come to life, showing you the signal's peaks and valleys, its rhythm and flow.

Waveforms: A Gallery of Electrical Art

Sine Waves: The graceful curves of alternating current (AC), like the ocean's tide.
Square Waves: The abrupt steps of digital signals, like a staircase to another dimension.
Triangle Waves: The sharp peaks and valleys of sawtooth patterns, like a mountain range.
Pulse Waves: The sudden bursts of energy, like a heartbeat or a drumbeat.
Complex Waves: The intricate patterns of sound and data, like a symphony or a coded message.

2. Measuring the Dance of Electricity

But the oscilloscope is more than just a visual tool. It's also a precise ruler, allowing you to measure the signal's vital stats:

Voltage: How high and low does the signal swing? The vertical scale gives you the answer, measured in volts per division.
Time: How long does it take for the signal to complete a cycle? The horizontal scale reveals the secret, measured in seconds per division.
With these measurements, you can unlock the signal's frequency, its pulse width, and other hidden characteristics. It's like taking the pulse of the electrical world!

Advanced Techniques: Unmasking the Signal's Secrets

Modern oscilloscopes are even more powerful, with features that automate measurements and reveal hidden details:

Automatic Measurements: Like a detective's toolkit, these functions instantly calculate voltage, frequency, and other key parameters.
Cursors: These on-screen markers pinpoint specific points on the waveform, allowing for precise measurements.
Math Functions: Perform calculations on the signal itself, like adding or subtracting different waveforms.
FFT Analysis: This tool reveals the signal's hidden frequencies, like a prism separating white light into its rainbow colors.

Case Studies: Real-World Oscilloscope Adventures

Automotive Diagnostics: A mechanic uses an oscilloscope to diagnose a misfiring engine, like a doctor listening to a heartbeat.
Telecommunications: An engineer uses an oscilloscope to ensure a clean and clear signal for a video conference, like a conductor tuning an orchestra.
Medical Research: A scientist uses an oscilloscope to study brainwaves, unlocking the secrets of human consciousness.

Conclusion: Your Journey into the World of Electrical Signals

The oscilloscope is your passport to the invisible world of electricity. It's a tool for discovery, measurement, and understanding. With it, you can unlock the secrets of electronic circuits, diagnose problems, and explore the fascinating dance of electrical signals. So, grab your probe, connect to the circuit, and embark on your oscilloscope adventure!

Nuclear physics

The nuclear model of the atom

The atom

Journey to the Heart of Matter: Unraveling the Atom's Secrets

Imagine a bustling city teeming with life, yet mostly composed of empty space. Buildings rise like colossal nuclei, housing the city's essence, while tiny, nimble electrons zip around like tireless delivery couriers, connecting everything together. This is the world of the atom, a universe in miniature.

Our understanding of this microscopic realm has undergone a dramatic evolution, from the ancient Greeks' concept of indivisible "Atomos" to the modern quantum mechanical model. Today, we know that atoms are not solid, indivisible spheres, but rather intricate structures composed of even smaller subatomic particles.

Delving into the Atom's Core

At the heart of the atom lies the nucleus, a dense, positively charged core containing protons and neutrons. These particles, collectively called nucleons, account for almost all of the atom's mass. Imagine the nucleus as the city's bustling downtown, where the majority of its inhabitants reside.

Surrounding the nucleus is a vast, cloud-like region where electrons reside. These negatively charged particles whiz around the nucleus like a swarm of bees, their movements governed by the laws of quantum mechanics. The electron cloud is enormous compared to the nucleus, making the atom mostly empty space. Think of it as the city's sprawling suburbs, where residents are spread out over a large area.

Key Concepts: Decoding the Atom's Language

To understand the atom's intricate structure, we need to familiarize ourselves with some key concepts:

Atomic Number (Z): The number of protons in an atom's nucleus. This is like the city's unique identification number, defining its character and properties.
Mass Number (A): The total number of protons and neutrons in the nucleus. This is analogous to the city's total population, including both permanent residents and temporary visitors.
Isotopes: Atoms of the same element (same atomic number) but with different numbers of neutrons (different mass numbers). These are like different neighborhoods within the city, each with its own unique characteristics.
Electron Shells and Subshells: Electrons occupy specific energy levels or shells around the nucleus, like apartments on different floors of a building. These shells are further divided into subshells, each with a unique shape and capacity for electrons, like individual rooms within an apartment.
Valence Electrons: The electrons in the outermost shell, which are involved in chemical bonding. These are like the city's ambassadors, interacting with other cities to form alliances and partnerships.

Alpha-Particle Scattering Experiments: Peering into the Atom's Depths

In the early 20th century, Ernest Rutherford and his colleagues conducted groundbreaking experiments that revolutionized our understanding of the atom. They fired alpha particles (positively charged particles consisting of two protons and two neutrons) at a thin gold foil and observed their scattering patterns.

Imagine throwing tennis balls at a fence. If the fence is solid, the balls will bounce back. But if the fence has large gaps, most balls will pass through. This is similar to what Rutherford observed.

Most alpha particles passed straight through the gold foil, suggesting that the atom is mostly empty space. However, a small fraction of alpha particles was deflected at large angles, some even bouncing back. This indicated the presence of a small, dense, and positively charged nucleus.

Rutherford's Atomic Model: A Miniature Solar System

Based on these experiments, Rutherford proposed a model of the atom that resembled a miniature solar system. The nucleus, like the Sun, occupies the center, while electrons, like planets, orbit around it.

This model was a major breakthrough, but it had limitations. It couldn't explain why electrons didn't spiral into the nucleus and why atoms emitted light at specific wavelengths.

Bohr Model: Quantized Energy Levels

Niels Bohr refined Rutherford's model by proposing that electrons orbit the nucleus in specific energy levels or shells. This explained why atoms emitted light at specific wavelengths, as electrons could only jump between these discrete energy levels.

Quantum Mechanical Model: Embracing Uncertainty

The current model of the atom, based on quantum mechanics, describes electrons as existing in a cloud-like region around the nucleus, their positions defined by probability distributions. This model embraces the wave-particle duality of matter and provides a more accurate description of atomic behavior.

Conclusion: The Atom's Enduring Mystery

The atom, once thought to be the smallest indivisible unit of matter, has revealed itself to be a complex and dynamic structure. From the ancient Greeks' "Atomos" to the modern quantum mechanical model, our understanding of the atom has evolved dramatically.

Yet, the atom continues to hold mysteries. The strong nuclear force that binds protons and neutrons together in the nucleus, the nature of dark matter and dark energy, and the ultimate fate of the universe are just some of the questions that continue to intrigue scientists.

As we delve deeper into the atom's secrets, we gain a greater appreciation for the intricate workings of the universe and our place within it. The journey to the heart of matter is a journey of discovery, revealing the beauty and complexity of the world around us.

The nucleus

The Heart of Matter: A Tiny Universe Within

Imagine a bustling city, teeming with life and activity. Now, shrink that city down to the size of a pinhead. That's the nucleus of an atom, a microscopic metropolis where protons and neutrons, the building blocks of matter, dance to the rhythm of the strong nuclear force.

Protons: The Identity Card of an Atom

Protons are like the city's mayor, holding a positive charge and dictating who's who in the atomic world. Each element has its own unique mayoral candidate – carbon always has six protons, oxygen eight, and so on. These proton mayors determine the city's population of electrons, the tiny workers that zip around the nucleus, shaping the city's character and how it interacts with its neighbors.

Neutrons: The Shape-Shifters

Neutrons are the city's quiet but influential residents. They don't carry any charge, but they add weight to the city, literally! Sometimes, a city might have a few extra neutrons or be missing a couple. These variations create isotopes, like sister cities with slightly different personalities. They might look the same, but their weight difference can lead to some unique behaviors.

The Strong Force: A Cosmic Glue

Holding this bustling city together is a force so powerful it makes gravity look like a weak handshake. It's the strong nuclear force, a cosmic glue that binds the protons and neutrons in a tight embrace, overcoming their natural tendency to repel each other. This force is the reason why the nucleus, despite its tiny size, is incredibly dense and stable. If you could scoop up a teaspoon of nuclear matter, it would weigh as much as a mountain!

The Nucleus in Action: From Ancient Secrets to Modern Marvels

This tiny metropolis isn't just a theoretical concept. It's the key to understanding some of the most fascinating phenomena in our world:

Whispers from the Past: Carbon-14, a radioactive isotope, is like a time capsule, allowing archaeologists to peer into the past. By measuring its decay, they can date ancient artifacts and uncover secrets hidden for millennia.

Healing Rays: Radioactive isotopes are also powerful tools in medicine. They can be used to target and destroy cancer cells or illuminate the inner workings of our bodies, helping doctors diagnose and treat diseases.

Powering the Future: Nuclear fission, the splitting of atoms, is a source of immense energy. While it comes with challenges, it also holds the potential to provide clean and abundant power for future generations.

Ion Formation: The Dance of Electrons

Atoms are usually electrically neutral, like a perfectly balanced scale. But sometimes, they lose or gain electrons, those tiny workers buzzing around the nucleus. This creates ions, atoms with an electrical charge.

Losing Electrons: The Birth of a Cation

When an atom loses electrons, it becomes a positive ion, or cation. It's like the city losing some of its workforce, leaving it with a net positive charge. Many factors influence this electron exodus, including the atom's energy levels and its desire to achieve a stable configuration, like a city striving for optimal efficiency.

Gaining Electrons: The Rise of an Anion

On the flip side, an atom can gain electrons, becoming a negative ion, or anion. This is like the city welcoming new workers, resulting in a net negative charge. Again, various factors play a role, including the atom's electron affinity and its quest for stability.

The Octet Rule: The Quest for Stability

Atoms have a strong desire to achieve a stable configuration of eight valence electrons, the outermost workers in the atomic city. This is known as the octet rule. It's like the city aiming for a perfect balance of workers to maintain order and prosperity.

Ions in Action: From Body to Battery

Ions are everywhere, playing vital roles in our world:

Electrolytes: The Spark of Life: These charged particles are essential for our bodies to function. They help transmit nerve impulses, regulate our heartbeat, and keep us hydrated.

Batteries: Power in the Palm of Your Hand: Batteries rely on the movement of ions to store and release energy, powering our phones, laptops, and even electric cars.

Corrosion: The Unwanted Guest: Corrosion, the gradual decay of metals, often involves the formation of ions. Think of the rust on an old car, a testament to the relentless dance of electrons.

A World of Wonder

The nucleus and the formation of ions are fundamental processes that shape the world around us. From the smallest atom to the largest star, these microscopic events drive the grand symphony of the universe. By understanding these concepts, we gain a deeper appreciation for the intricate beauty and complexity of matter, the very essence of our existence.

Imagine...

You're not just reading about atoms, you're shrinking down to an impossible size, zooming through the vast emptiness of an atom to land smack-dab in the middle of its bustling nucleus. Forget boring textbooks, this is a subatomic safari!

Meet the Locals:

Protons: These guys are like the nucleus's identity cards. Each element has its own unique proton posse – carbon rocks 6, oxygen flaunts 8, and so on. They're also the reason atoms are electrically neutral, since they balance out the negatively charged electrons zipping around outside. Think of them as the anchors of the atom.
Neutrons: These are the proton's chill buddies, hanging out in the nucleus and adding to the atom's mass. Protons and neutrons are basically the "mass crew" of the atom, like the heavy lifters in a weightlifting competition.

Nucleon Number (A): The Atom's Weight Class

This number tells you the total number of protons and neutrons crammed into the nucleus. It's like weighing the whole nucleus together – the bigger the number, the heavier the atom.

Nuclide Notation: The Atom's Secret Code

Each atom has its own secret code, like a microscopic fingerprint. It looks like this:

^A_ZX

X: The element's symbol (like C for carbon, O for oxygen).
A: The nucleon number (the "weight class").
Z: The proton number (the "identity card").
With this code, you can instantly tell what kind of atom you're dealing with.

Isotopes: The Atom's Quirky Cousins

Imagine atoms with the same name (same element) but slightly different weights. That's what isotopes are! They have the same number of protons (same identity) but different numbers of neutrons (different weights).

Think of it like this:

Carbon-12: The "normal" carbon, like the one in your pencil.
Carbon-14: Carbon's radioactive cousin, used to date ancient artifacts. It's like carbon-12, but with a couple extra neutrons making it a bit unstable and prone to emitting energy.

Isotopes in Action: More Than Just Atomic Oddities

These atomic variations aren't just theoretical curiosities. They're powerful tools used in:

Medicine: Radioactive isotopes help doctors see inside the body and fight diseases like cancer.
Archaeology: Carbon-14 helps archaeologists date ancient relics and piece together the past.
Climate Research: Scientists analyze isotopes in ancient ice and sediments to understand how Earth's climate has changed over time.

The Atomic World: A Universe of Wonder

Understanding atoms is like unlocking a hidden universe within our own. By grasping the roles of protons, neutrons, and isotopes, we can better appreciate the complexity and beauty of the world around us – and even harness the power of atoms for the benefit of humankind.

Radioactivity
Detection of radioactivity

Unlocking the Secrets of Invisible Rays: A Journey into Radiation Detection

Imagine a world where we could see the invisible dance of energy, where the whispers of atoms become visible trails of light and sparks. That's the magic of radiation detection, a field that allows us to peer into the hidden realm of alpha, beta, and gamma radiation.

1. The Cloud Chamber: Painting with Radioactive Brushstrokes

Step into a misty chamber, a realm of supersaturated vapor hanging heavy with anticipation. This is the cloud chamber, a time machine that takes us back to the early days of radiation research.

Alpha Particles: The Bold Strokes - Like a cosmic artist wielding a thick brush, alpha particles leave behind short, bold streaks, their tracks a testament to their hefty mass and powerful ionizing force.
Beta Particles: The Wispy Lines - Nimble and light, beta particles weave a more delicate tapestry, their paths a testament to their erratic dance through the vapor.
Gamma Rays: The Elusive Ghosts - Gamma rays, the ninjas of the radiation world, leave no direct trace. Yet, their presence is revealed by the faint trails of the electrons they knock loose, like ghostly whispers in the chamber.

2. The Spark Counter: A Symphony of Sparks

Imagine a miniature lightning storm, where each spark tells a story. That's the spark counter, a simple yet elegant device that captures the drama of alpha particles.

Alpha Particles: Igniting the Spark - As an alpha particle zip through the air, it sets off a chain reaction, a cascade of ionization that culminates in a bright spark, a tiny firework announcing its arrival.

3. The Geiger-Müller Tube: The Radiation Whisperer

Enter the Geiger-Müller tube, a sentinel listening to the whispers of radiation. This versatile device, a workhorse of the radiation detection world, counts the silent footsteps of alpha, beta, and gamma radiation.

Beta and Gamma Radiation: The Clickety-Clack of Counts - Each click of the Geiger counter is a testament to the passage of a beta or gamma ray, a tiny pulse of current amplified into an audible signal.

4. Beyond the Classics: The New Detectives on the Block

In the ever-evolving world of radiation detection, new heroes have emerged, armed with cutting-edge technology:

Scintillation Detectors: The Light Catchers - These detectors capture the faint glimmers of light emitted when radiation strikes certain materials, translating these luminous whispers into electrical signals.
Solid-State Detectors: The Electron Counters - These miniature marvels use semiconductors to directly count the electrons knocked loose by radiation, offering unparalleled precision and efficiency.
Imaging Detectors: The Radiation Photographers - These advanced devices create stunning images of radiation's distribution, providing invaluable insights in fields ranging from medical imaging to security.

Conclusion: Unveiling the Invisible

From the ethereal beauty of cloud chambers to the precise clicks of Geiger counters, the world of radiation detection is a testament to human ingenuity. These tools, both classic and cutting-edge, allow us to unravel the mysteries of invisible rays, ensuring our safety and unlocking new frontiers of knowledge.

The Unseen Symphony: A Journey into Background Radiation
Imagine a world humming with invisible energy, a constant symphony of particles dancing around us. This is the world of background radiation, a subtle yet pervasive force that has shaped life on Earth since its very beginning.

1. The Silent Concert: What is Background Radiation?

We live in a radioactive world. Not in the way of glowing green goo and superhero origin stories, but in a much more subtle way. Background radiation is like a gentle rain of energy, constantly falling on us from the cosmos and the Earth itself. It's a natural phenomenon, a whisper of the universe's creation and the ongoing decay of elements within our planet. This "radiation" is simply energy traveling in the form of waves or particles. While it can interact with our bodies and even cause harm in high doses, our cells have evolved remarkable mechanisms to repair most of the damage from these low levels of natural radiation.

2. The Main Players: Where Does this Energy Come From?

Think of background radiation as an orchestra, with different instruments contributing to the overall sound. Here are some of the key players:

Radon: The Stealthy Maestro - This colorless, odorless gas seeps from the Earth's crust, a silent reminder of the radioactive decay happening beneath our feet. Radon can accumulate in homes, and while it's harmless in low concentrations, prolonged exposure can increase the risk of lung cancer. It's like a bassoon in our symphony, its low tones potentially harmful if they become too loud.

Rocks and Buildings: The Earthy Chorus – Even the very ground we walk on and the buildings we inhabit contribute to the radiation symphony. Granite countertops, brick walls, concrete foundations – all contain trace amounts of naturally occurring radioactive materials. Think of them as the percussion section, adding a steady beat to the background radiation.

Food and Drink: The Nourishing Interlude - Believe it or not, even the food we eat and the water we drink contain minute amounts of radioactive isotopes. Potassium-40, a naturally occurring radioactive isotope of potassium, is found in bananas, potatoes, and even us! This is like the gentle flute in our orchestra, a subtle but essential part of the composition.

Cosmic Rays: The Celestial Soloist - From the vast expanse of space, high-energy particles rain down upon us, remnants of exploding stars and solar flares. The higher we go, the less atmosphere there is to shield us, which is why airline pilots and frequent flyers receive a slightly higher dose of radiation. Imagine this as a soaring violin solo, its intensity increasing as we ascend to greater heights.

3. Tuning Out the Noise: Measuring Radiation

Scientists use specialized instruments to "listen" to this radiation symphony. But how do they isolate the sound of a specific source when surrounded by the constant hum of background radiation? It's like trying to hear a single conversation in a crowded room.

To do this, they measure the "count rate" – the number of radiation events detected per unit of time. By first measuring the background count rate in the absence of the source they're interested in, they can then subtract this "noise" from the total count rate when the source is present. This allows them to accurately determine the radiation emitted solely by the source, like isolating the melody from the rest of the orchestra.

Case Study: A Radioactive Iodine Serenade

Imagine a researcher studying the radioactivity of iodine-131, a medical isotope used in treating thyroid conditions. Their radiation detector picks up a total count rate of 500 counts per minute. However, they know that background radiation in the lab contributes 30. By subtracting this background "noise," they find the true count rate of the iodine-131 sample to be 470.

The Grand Finale

Background radiation is an inescapable part of our existence, a constant reminder of the dynamic universe we inhabit. While it might seem a bit daunting, understanding its sources and levels allows us to appreciate its role in the natural world and take appropriate precautions when necessary. So, the next time you enjoy a banana or gaze at the stars, remember the unseen symphony playing all around you, a testament to the intricate and awe-inspiring nature of our universe.

The three types of emission

Imagine the atom as a bustling city...

...filled with protons and neutrons, all living together in the nucleus, the city center. But sometimes, this city gets overcrowded and unstable. Think of it like a party that's gotten a little too wild! To regain some peace and quiet, the atom needs to release some energy. That's where radioactive decay comes in – it's the atom's way of letting off some steam.

Alpha Decay: The Helium Balloon Escape

Picture a helium balloon escaping from the party. That's alpha decay! A cluster of two protons and two neutrons (like a tiny helium nucleus) decide to make a break for it. They're big and a bit clumsy, bumping into things as they go. They don't get far, but they leave a trail of chaos (ionization) in their wake. Think of them as the partygoers who stumble into furniture and knock over drinks!

Beta Decay: The Shape-Shifting Electron

Now imagine one of the partygoers suddenly transforming into a completely different person – that's beta decay! A neutron in the nucleus decides it's had enough and morphs into a proton, shooting out an electron (or its quirky antimatter twin, the positron) in the process. It's like a magical quick-change act, altering the atom's identity.

Gamma Decay: The After-Party Chill

After all the excitement of alpha and beta decay, the atom needs to relax and unwind. That's where gamma decay comes in. It's like the soothing music playing as everyone chills out after the party. The atom releases its remaining energy as gamma rays, invisible waves of energy that calm things down.

Radioactive Decay: A Tale of Three Transformations

Each type of decay is unique, like different characters in our atomic city's story:

Alpha decay: The clumsy helium balloon escape, leaving a trail of ionization.
Beta decay: The shape-shifting electron, changing the atom's identity.
Gamma decay: The after-party chill, releasing calming energy waves.

Real-World Adventures

These atomic tales aren't just theoretical – they play out in the real world all the time!

Alpha particles are the heroes in smoke detectors, sensing danger and sounding the alarm.
Beta particles help doctors diagnose and treat diseases, acting like tiny spies inside the body.
Gamma rays are the powerful superheroes of medical imaging, revealing hidden secrets within the human body.

The Unpredictable Dance

But here's the twist – radioactive decay is completely random and spontaneous. It's like a dance floor where the atoms move unpredictably, emitting radiation in all directions. We can't predict when or where it will happen, adding a touch of mystery to this atomic adventure.

The Enduring Fascination

Radioactive decay is a captivating phenomenon that continues to intrigue scientists and shape our world. From medical breakthroughs to industrial applications, it's a reminder of the incredible power hidden within the tiniest building blocks of matter. So next time you think about atoms, remember the bustling city, the escaping balloons, the shape-shifting electrons, and the calming gamma rays – a microscopic world of constant transformation and endless fascination.

A Quirky Guide to Atomic Adventures: Alpha, Beta, and Gamma Radiation

Buckle up, science enthusiasts! We're about to embark on a thrilling quest into the wild world of radiation, where atoms get a bit rowdy and particles fly like popcorn in a microwave. Today's spotlight shines on the mischievous trio: Alpha, Beta, and Gamma radiation.

1. The Ionization Circus and the Great Escape Act

a) Ionization: Stripping Atoms Bare!

Imagine atoms as tiny, bustling cities, with electrons as their busy citizens. Now, picture radiation as a mischievous imp with a knack for causing chaos. When this imp zooms through an atom-city, it can knock off some electrons, leaving the atom a bit "undressed" and turning it into what scientists call an ion.

Alpha (α): The Heavyweight Champion
Alpha particles are like burly Sumo wrestlers – big, positively charged, and packed with energy. They barge through atom-cities, leaving a trail of ionized atoms in their wake. This makes them the ultimate ionization brawlers, but their bulk also means they tire quickly and don't travel far. Think of them as the life of the party, but only for a short while.

Beta (β): The Speedy Sprinter
Beta particles are like nimble ninjas – small, negatively charged, and lightning-fast. They zip through atom-cities, occasionally knocking off an electron or two. They're less disruptive than alpha particles but can cover more ground before calling it a day.

Gamma (γ): The Ghostly Glide
Gamma rays are like ethereal phantoms – no charge, no mass, just pure energy. They waft through atom-cities, rarely interacting with the residents. They're the ultimate escape artists, penetrating deep into matter without leaving much of a trace.

Example Time!

Imagine a playground with a sandpit (our atom-city). Alpha particles are like rambunctious toddlers throwing sand everywhere (high ionization). Beta particles are like kids on scooters, zipping around and occasionally kicking up some sand (moderate ionization). Gamma rays are like stealthy drones hovering above, barely disturbing the sand below (low ionization).

Case in Point: Chernobyl's Triple Threat

The Chernobyl disaster unleashed a radioactive rampage, showcasing the distinct personalities of our radiation trio. Alpha particles, the party animals, mostly stayed confined within the reactor building. Beta particles, the sprinters, ventured further, causing localized havoc. Gamma rays, the escape artists, went on a global tour, prompting widespread evacuations and long-term monitoring.

b) Penetrating Power: The Great Wall of Matter

Now, let's see how our radiation buddies fare against barriers.

Alpha (α): The Paper Tiger
Alpha particles are easily stopped by a mere sheet of paper or a few centimeters of air. They're like fierce warriors with a weakness for flimsy fences.

Beta (β): The Aluminum Assassin
Beta particles can penetrate further, breezing through air and even thin sheets of aluminum. They're like skilled archers whose arrows can pierce moderate defenses.

Gamma (γ): The Lead Leviathan
Gamma rays are the ultimate penetration masters, requiring thick lead shields or concrete barriers to stop them. They're like unstoppable siege engines, capable of breaching even the strongest fortresses.

Example Time!

Think of a castle under siege. Alpha particles are like foot soldiers easily repelled by the outer walls. Beta particles are like archers whose arrows can reach the inner courtyard. Gamma rays are like powerful catapults that can hurl projectiles over the walls and into the heart of the castle.

Case in Point: Medical Imaging's Invisible Eye

Gamma rays' penetrating prowess makes them invaluable in medical imaging. Techniques like PET scans and gamma cameras harness their ability to traverse the body, providing a glimpse into our inner workings.

2. Electric and Magnetic Fields: The Particle Dance Party

Charged particles, like alpha and beta, can't resist the allure of electric and magnetic fields. They're like tiny dancers, twirling and swirling under the influence of these invisible forces.

Electric Fields: The Opposites Attract Tango
Alpha particles, being positively charged, shimmy towards the negative plate in an electric field. Beta particles, with their negative charge, groove towards the positive plate. Gamma rays, having no charge, remain aloof, like wallflowers at the dance party.

Magnetic Fields: The Left-Hand Rule Boogie
Alpha and beta particles do a synchronized dance in magnetic fields, deflecting in opposite directions due to their opposing charges. The direction of their moves can be predicted using the "Left-Hand Rule" – a handy trick for physics aficionados. Gamma rays, again, sit this one out, unfazed by the magnetic rhythm.

Example Time!

Imagine a disco ball casting mesmerizing patterns (electric and magnetic fields). Alpha and beta particles are like dancers moving in opposite directions under the dazzling lights. Gamma rays are like the DJ, unaffected by the light show but controlling the vibe.

Case in Point: The Large Hadron Collider's Particle Ballet

The Large Hadron Collider, a gigantic particle accelerator, uses powerful magnetic fields to choreograph the movements of charged particles. This allows scientists to study their properties and interactions, unlocking the secrets of the universe.

Bonus Round!

Remember, the energy of beta and gamma radiation also influences their penetrating power and deflection. Higher energy means greater penetration and less deflection. And don't forget about scattering, where radiation bounces off matter like a hyperactive bouncy ball, changing its direction and energy.

The Grand Finale

Alpha, beta, and gamma radiation, each with its unique quirks and talents, play vital roles in various fields, from medical treatments to industrial applications. Understanding their properties is key to harnessing their power and ensuring radiation safety. So, keep exploring the atomic realm, and who knows what other wonders you might uncover!

Radioactive decay

Imagine the atom as a bustling city...

...filled with protons and neutrons, like citizens, residing in the central nucleus, the city's vibrant downtown. Most cities hum along peacefully, but some have an unstable energy, a tension in the air. These are our "radionuclides." To find balance, they undergo a transformation, a "radioactive decay," releasing energy like a burst of fireworks, or a captivating light show, changing the very character of the city.

Think of it like this:

Alpha decay? It's like a helium-filled balloon escaping the city, drifting away with a couple of protons and neutrons, leaving the city a bit smaller and less crowded.
Beta decay? Imagine a tiny, speedy electron, like a courier, zipping out of the city carrying a message, transforming a neutron into a proton and changing the city's identity.
Gamma decay? This is pure energy, a brilliant flash of light, like a beacon signaling the city's transformation, a release of excitement after some internal shifts.
Decay equations? They're like city records, documenting the changes, the comings and goings of these subatomic citizens. Each element has its own unique symbol, like a city's flag, with numbers indicating its population of protons and neutrons.

The half-life? It's like predicting when half the city's population might move out. It's a mystery, a game of chance, we know it will happen, but the exact timing remains elusive.

But here's where it gets really interesting...

Radioactive decay isn't just some abstract concept. It's the driving force behind some of the most incredible applications:

Medicine: Imagine tiny radioactive tracers, like miniature detectives, illuminating the body's hidden corners, helping doctors diagnose and treat diseases.
Archaeology: Like a time, machine, radioactive decay helps us unravel the mysteries of the past, dating ancient artifacts and revealing secrets buried for millennia.
Industry: It's like having X-ray vision, allowing us to peer inside machines and structures, ensuring safety and efficiency.
Nuclear Power: Harnessing the immense energy within atoms, like capturing the power of a thousand suns, to generate electricity for our modern world.

But with great power comes great responsibility...

The Chernobyl disaster, a stark reminder of the destructive potential when things go wrong, a city overwhelmed by an uncontrolled release of energy. The Goiania accident, a cautionary tale of radioactive materials falling into the wrong hands, highlighting the need for careful handling and respect for this potent force.

Radioactive decay is a story of transformation, a dance of energy and matter, a reminder of the constant change that shapes our world. It's a story that continues to unfold, revealing new secrets and possibilities with each passing day.

Fission and fusion

Nuclear Fusion: A Star's Dance of Energy

Imagine the heart of a star, a cosmic cauldron where atoms tango to a rhythm of unimaginable heat and pressure. This is nuclear fusion, the universe's own power plant, where lighter atomic nuclei, like shy dancers, overcome their initial repulsion and twirl into a passionate embrace, forming a heavier nucleus and releasing a dazzling burst of energy.

The Fusion Tango: A Step-by-Step Guide

Setting the Stage (Extreme Temperatures and Pressures): The dance floor heats up to millions of degrees Celsius, and the pressure intensifies, mimicking the crowded heart of a star. This intense environment forces the positively charged nuclei, initially hesitant like wallflowers, to overcome their electrostatic repulsion.

Breaking the Ice (Overcoming the Coulomb Barrier): The nuclei, now energized by the heat, gain the confidence to approach each other. They move with incredible speed, finally breaking through the "Coulomb barrier," the invisible wall of electrostatic force that keeps them apart.

The Quantum Leap (Tunnel Effect): Even with their newfound confidence, not all nuclei possess the energy to completely break through the barrier. But like a secret passageway, quantum mechanics allows some to "tunnel" through, defying classical physics and joining the dance.

The Grand Finale (Nuclear Binding Energy and Mass Defect): As the nuclei merge in a final, passionate embrace, a tiny portion of their mass transforms into pure energy, a burst of light and heat, as dictated by Einstein's famous $E=mc^2$. This mass, seemingly vanished, is the "mass defect," and the energy released is the "nuclear binding energy," the grand prize of this cosmic dance.

Fusion in Stars: The Proton-Proton Waltz

In stars like our Sun, the primary fusion dance is the "proton-proton chain reaction," where hydrogen nuclei (protons) waltz to form helium. It's a three-step routine:

Two Protons, One Deuterium: Two protons, the shyest dancers, take the first step, fusing to form deuterium, a heavier hydrogen isotope. This delicate step involves the "weak nuclear force" and the release of a positron and a neutrino, like whispers of encouragement from the cosmic audience.

Deuterium and a Proton, a Helium-3 Trio: Deuterium, now more confident, joins another proton, forming helium-3, and releasing a gamma-ray photon, a flash of light announcing their success.

Two Helium-3s, a Helium-4 Finale: Two helium-3 nuclei, the stars of the show, perform the final, dazzling twirl, merging into helium-4 and releasing two protons, ready to join the next dance.

This captivating performance, repeated countless times, is what fuels the stars, their radiant energy a testament to the power of fusion.

Earthly Fusion: Mirroring the Stars

While we marvel at the stars' energy, scientists on Earth are striving to recreate this cosmic dance, hoping to harness its power for clean energy. It's a challenging endeavor, like teaching humans to dance like stars:

Magnetic Confinement Fusion: Imagine a magnetic bottle, where powerful magnetic fields, like invisible walls, confine and control the hot plasma, the dance floor for fusion. Tokamaks and stellarators are examples of these intricate "dance halls."

Inertial Confinement Fusion: Here, high-powered lasers or particle beams act as spotlights, compressing and heating a tiny pellet of fusion fuel, igniting a miniature star on Earth.

Despite the challenges, progress is being made. ITER, a massive international collaboration, is building a tokamak in France, hoping to achieve net energy gain, a standing ovation for human ingenuity. The National Ignition Facility (NIF) in the US has already achieved "ignition," a moment where the fusion reaction generates more energy than the lasers used to initiate it, a critical milestone in this ambitious performance.

Nuclear Fission: Splitting the Atom

While fusion unites, nuclear fission divides. It's a process where a heavy atomic nucleus, like an unstable dancer, splits into smaller fragments, releasing energy in the process. This is the power behind nuclear power plants and atomic bombs.

The Fission Breakdown

Neutron Absorption: A heavy nucleus, like uranium-235, absorbs a neutron, an unwelcome guest that disrupts its stability.

Nuclear Instability: The nucleus, now agitated, begins to deform, its internal forces struggling to maintain balance.

Splitting the Nucleus: Like a dancer losing their footing, the nucleus splits into smaller fragments, called fission products.

Release of Energy and Neutrons: These dramatic split releases a surge of energy and several neutrons, like shrapnel from the breakup.

Chain Reactions: A Cascade of Splits

The released neutrons can trigger further fission reactions in other nuclei, creating a chain reaction, a cascade of splits.

Controlled Chain Reactions: In nuclear reactors, control rods and moderators act as careful choreographers, managing the chain reaction, ensuring a steady, controlled release of energy, a graceful, predictable dance.

Uncontrolled Chain Reactions: In atomic bombs, the chain reaction is unleashed, an uncontrolled frenzy of splits, resulting in a devastating explosion.

Fission vs. Fusion: A Tale of Two Dances

Feature	Fission	Fusion
Process	Splitting a heavy nucleus	Combining light nuclei
Fuel	Uranium, plutonium	Hydrogen isotopes
Products	Fission products, neutrons	Helium, neutrons
Energy release	Lower	Higher
Radioactive waste	Significant	Minimal
Technological maturity	Well-established	Under development
Safety concerns	Potential for meltdowns	Less risk of meltdowns

Conclusion: The Future of Energy

Both fission and fusion, each a unique dance of the atom, offer powerful energy sources. Fission, a mature technology, powers our current nuclear plants. Fusion, the star's own energy source, remains a tantalizing dream, a challenging choreography that promises a cleaner, more abundant energy future. As we continue to explore these atomic dances, we inch closer to harnessing the fundamental forces of the universe, shaping a future where energy is as boundless as the stars themselves.

Half-life

Imagine a clock that ticks with the rhythm of decaying atoms...

That's essentially what half-life is! It's the time it takes for half of a radioactive substance to disappear, like a ghostly hourglass marking time in the atomic world.

Think of it like this: you have a bag of popcorn kernels, and every minute, half of them magically pop. That's half-life in action! But instead of popcorn, we're talking about tiny, unstable atoms that transform into something new.

Why should you care about half-life?

Well, it's not just a science quirk. This atomic rhythm helps us:

Unmask the age of ancient treasures: Like a time-traveling detective, radiocarbon dating uses the half-life of Carbon-14 to reveal the age of fossils, ancient scrolls, and even the Shroud of Turin!
Peer into the human body: Doctors use radioactive isotopes with specific half-lives to create amazing images of our insides and even target pesky tumors.
Track invisible pollutants: Half-life helps us follow the trail of radioactive materials in the environment, like tiny radioactive breadcrumbs.

Speaking of radiocarbon dating...

Imagine you're an archaeologist, unearthing a mysterious wooden artifact. How can you tell how old it is? Enter radiocarbon dating, the Sherlock Holmes of the archaeology world!

While alive, plants and animals absorb Carbon-14, a special kind of carbon created in the atmosphere. When they die, the Carbon-14 clock starts ticking, decaying at a steady rate. By measuring how much Carbon-14 is left, we can rewind the clock and figure out when that organism met its end.

But it's not always that simple. Imagine trying to read a clock that's been buried in the dirt for centuries! Scientists have to carefully clean and calibrate their samples, accounting for things like changes in the atmosphere over time.

The future of atomic clocks:

Scientists are always pushing the boundaries of what's possible. They're refining techniques, shrinking instruments, and even using statistics to get even more precise dates. Who knows what mysteries we'll unlock next with the help of these atomic clocks?

So, next time you hear about half-life, remember: it's not just a scientific concept, it's a key to unlocking the secrets of time, hidden within the very fabric of matter.

Radioactive Isotopes: The Tiny Superheroes with a Split Personality

They might sound like something out of a sci-fi comic book, but radioactive isotopes are actually real-life superheroes with a secret identity. They're the silent forces behind everyday objects like smoke detectors, and they even play a starring role in keeping our food safe and our hospitals sterile. But like any superhero, they have their quirks – namely, the kind of "rays" they shoot out and how long their superpowers last.

The "Fantastic Four" of radiation

Think of radioactive isotopes as having different superpowers in the form of radiation. There are four main types:

Alpha Particles: These are like tiny, burly bouncers. They're strong but can't reach very far, easily stopped by a thin barrier (like a sheet of paper!). Not the best for fighting crime unless it's happening right under their noses.

Beta Particles: These are like speedy ninjas, faster and with more range than the bouncers. They can zip through the air and even penetrate thin materials. Pretty cool, right?

Gamma Rays: Now these are the real heavy hitters, like Superman's heat vision! They can travel long distances and penetrate even thick walls. Definitely the ones you want on your side in a radioactive showdown.
Neutron Radiation: Imagine these as shape-shifters, able to change the very nature of the atoms they encounter. They're powerful but a bit unpredictable, so they need to be handled with extra care.

Half-Life: The Isotope's "Use-By" Date

Every superhero has their limits, and for radioactive isotopes, that's their half-life. It's like a countdown timer for their powers. Some isotopes are like energetic puppies, burning bright and fast with a short half-life. Others are like wise old sages, radiating their energy slowly and steadily over a long half-life.

Now, let's see these radioactive superheroes in action!

(a) Smoke Detectors: Americium-241 to the Rescue!

Imagine a tiny superhero, Americium-241, standing guard in your smoke detector. It uses its gamma ray vision to constantly scan for smoke. When those pesky smoke particles try to sneak by, Americium-241 raises the alarm, giving you time to escape! And with a half-life of over 400 years, it's like having a superhero with a lifetime contract!

(b) Food Safety: Zapping Bacteria with Gamma Rays

Ever wonder how your favorite foods stay fresh for so long? It's thanks to our gamma ray superheroes like Cobalt-60 and Cesium-137! They blast those nasty bacteria with their powerful rays, keeping your food safe and delicious. And because they have relatively short half-lives, they disappear quickly, leaving no radioactive residue behind.

(c) Sterilization: Gamma Rays to the Rescue (Again!)

In hospitals and labs, gamma ray superheroes like Cobalt-60 are hard at work sterilizing equipment, making sure it's safe for doctors and scientists to use. Their penetrating rays reach every nook and cranny, wiping out even the toughest germs.

The End (or is it?)

So, there you have it! Radioactive isotopes, the unsung heroes of our everyday lives. They might sound a bit scary, but with the right understanding and careful handling, they're powerful allies in making our world a safer, healthier place.

Imagine a world where we can see through solid objects, where tiny radioactive superheroes battle cancer within our bodies, and where factories hum with the precision of invisible rays. This isn't science fiction; it's the fascinating reality of how we harness the power of radioactive isotopes.

These isotopes, with their unique ability to emit radiation, are like microscopic beacons, each with its own superpower depending on the type of radiation it emits and how quickly it fades away (its half-life). Let's delve into this world and explore how these tiny powerhouses help us control the thickness of materials and even combat cancer.

Radioactive Isotopes: The Ultimate Quality Control Inspectors

Imagine a paper mill churning out endless reams of paper. How do they ensure each sheet is precisely the same thickness? Enter our radioactive superheroes, emitting beta radiation, like tiny inspectors with X-ray vision. They can see through the paper, measuring its thickness with incredible accuracy. If the paper is too thick or too thin, they instantly alert the machines to adjust, ensuring every sheet is perfect.

These isotopes are chosen carefully. Alpha radiation, for instance, is like a clumsy giant – powerful but easily stopped, suitable only for the thinnest materials. Beta radiation, on the other hand, is like a nimble acrobat, perfect for gauging thin to medium-thickness materials. And for the thickest materials, like steel plates, we need the incredible penetrating power of gamma rays, like microscopic lasers.

The Half-Life Balancing Act

The half-life of an isotope is like its stamina. Some isotopes are sprinters, decaying quickly and needing frequent replacement, while others are marathon runners, providing a stable source of radiation for a long time. Choosing the right isotope is a delicate balance – we need enough power to get the job done but not so much that it becomes a safety hazard.

Radioactive Isotopes: The Cancer-Fighting Superheroes

Now, let's journey into the human body, where radioactive isotopes take on a new role as cancer-fighting superheroes. They team up with special molecules that act like homing beacons, guiding them directly to the cancerous cells.

In diagnosis, these isotopes emit gamma rays, like tiny signal flares, allowing doctors to pinpoint the location and size of the tumor. It's like having a microscopic spy inside the body, reporting back on the enemy's position.

In treatment, these isotopes become tiny warriors, delivering a targeted blow to the cancer cells with their radiation. They damage the DNA of these rogue cells, halting their growth or even destroying them completely.

Choosing the Right Hero for the Job

Just as in thickness control, the type of radiation and half-life are crucial in cancer treatment. For diagnosis, we need isotopes with short half-lives to minimize radiation exposure to the patient, while for treatment, we might choose isotopes with longer half-lives to provide a continuous attack on the cancer cells.

The Future of Radioactive Isotopes

The story of radioactive isotopes is still being written. Scientists are constantly discovering new ways to harness their power, developing new isotopes with even more specific abilities. From creating new materials to exploring the mysteries of the universe, these tiny powerhouses are shaping our world in ways we are only beginning to understand.

So, the next time you hear the word "radioactive," don't just think of danger. Think of the microscopic superheroes working tirelessly behind the scenes, ensuring the quality of our products, diagnosing and treating diseases, and pushing the boundaries of human knowledge. They are a testament to the incredible power of science and the endless possibilities that lie within the atom.

Safety precautions

The Unseen Sculptor: A Tale of Energy, Atoms, and Life (2024)

Imagine an invisible artist, a sculptor of the unseen, with the power to reshape the very building blocks of life. This artist is ionizing radiation, a form of energy released by atoms, capable of altering the delicate balance within living cells. Like a master craftsman, it can chip away at the intricate structures of life, leaving behind a trail of both subtle and profound changes.

A Dance of Disruption

Ionizing radiation, in its various forms – alpha particles, beta particles, gamma rays, and X-rays – is like a dancer with varying steps and rhythms. Each type of radiation interacts with living things in its unique way, leaving its signature on the dance floor of life.

Direct Action: A Close Embrace

Sometimes, the dancer engages in a close embrace with the DNA molecules, the carriers of life's blueprint. This direct interaction can break the delicate bonds holding the DNA strands together, like a passionate dance move gone awry. Single-strand breaks are like minor missteps, often repaired with ease. But double-strand breaks are more like a dramatic fall, requiring intricate repair work to restore the dance floor.

Indirect Action: A Ripple Effect

Other times, the dancer's movements create a ripple effect, interacting with the water molecules that surround the DNA. This indirect action generates free radicals, like mischievous sprites that dart around, causing damage to the DNA and other cellular components. These sprites, with their fleeting existence, can leave a lasting impact on the dance of life.

The Intensity of the Dance

The impact of the dance depends on several factors:

The Dancer's Energy: Different types of radiation, like dancers with varying levels of energy, have different impacts. Alpha particles, with their heavy steps, cause localized damage, while gamma rays, with their energetic leaps, can affect a wider area.

The Duration of the Dance: The longer the dance, the more significant the impact. Higher doses of radiation, like a prolonged dance session, can lead to more severe consequences.

The Tempo of the Dance: A fast-paced dance can be more disruptive than a slow and graceful one. A high dose rate, like a rapid tempo, gives the cells less time to recover and repair the damage.

The Dance Floor: Different tissues and organs, like different dance floors, have varying sensitivities to the dance of radiation. Rapidly dividing cells, like those in the bone marrow, intestines, and reproductive organs, are more susceptible to the disruptive moves.

The Dancer's Partner: Each individual, like a unique dance partner, responds differently to the dance of radiation. Genetic factors and overall health can influence the impact of the dance.

The Aftermath of the Dance

The dance of ionizing radiation can leave behind a trail of consequences:

Cell Death: Sometimes, the damage is too severe, and the dancer's partner, the cell, succumbs to the intensity of the dance. Apoptosis, like a graceful exit, is a programmed cell death that eliminates irreparably damaged cells. Necrosis, on the other hand, is a more chaotic end, with the cell swelling and rupturing, releasing its contents and causing further disruption. Mitotic catastrophe occurs when a damaged cell attempts to divide, leading to a chaotic and often fatal outcome.

Mutations: A Change in the Choreography: When the DNA damage is not repaired correctly, it can lead to mutations, like a change in the choreography of life. These mutations can be silent, like a subtle change in the music, or they can be more dramatic, altering the very steps of the dance. Mutations in somatic cells can contribute to cancer development, while mutations in germ cells can be passed on to future generations, like a legacy of the dance.

Cancer: A Tumultuous Dance: One of the most significant long-term effects of the dance of ionizing radiation is an increased risk of cancer. Radiation can initiate or promote the growth of tumors, like a chaotic and uncontrolled dance that disrupts the harmony of the body.

Lessons from the Dance Floor

Throughout history, there have been instances where human populations were exposed to significant levels of ionizing radiation, providing valuable lessons about the impact of this invisible dancer. The atomic bombings of Hiroshima and Nagasaki, the Chernobyl disaster, and medical radiation exposure have all contributed to our understanding of the dance of radiation and its consequences.

The Future of the Dance

Ongoing research continues to shed light on the intricate dance of ionizing radiation and its effects on living things. Scientists are exploring the impact of low doses, developing individualized risk assessment models, and advancing radiation protection and mitigation strategies.

Embracing the Knowledge

By understanding the dance of ionizing radiation, we can make informed decisions about our exposure and contribute to a safer and healthier future. Like an informed audience, we can appreciate the power and potential of this invisible dancer while respecting its ability to shape the very essence of life.

Radioactive Rendezvous: A Journey into the Heart of Nuclear Safety

Imagine a world where atoms dance with an invisible energy, a world where substances hold the power to heal and harm in equal measure. This is the realm of radioactive materials, where the stakes are high and safety is paramount. Join us on a thrilling expedition into the fascinating world of radioactive materials, where we'll unravel the secrets of their safe handling, movement, and storage.

Act I: The Radioactive Road Trip

Our journey begins with the transportation of these enigmatic materials, a process akin to a high-stakes escort mission. Picture specially designed containers, resembling armored vehicles for atoms, shielding their potent cargo from the outside world. These containers are the unsung heroes of nuclear safety, ensuring that radioactive materials reach their destination without causing a commotion.

The movement of these materials is a carefully choreographed ballet, with dedicated vehicles gliding along meticulously planned routes, like seasoned performers on a grand stage. Trained personnel, the guardians of this invisible energy, navigate these routes with precision and expertise, ensuring that the show goes on without a hitch.

Act II: The Radioactive Stage

Our next stop takes us to the diverse settings where radioactive materials take center stage – hospitals, industrial plants, and research labs. In hospitals, these materials transform into healing agents, aiding in the diagnosis and treatment of diseases. Industrial plants harness their power for gauging and monitoring, while research labs explore their mysteries to unlock the secrets of the universe.

In each of these settings, safety protocols are the choreographers, dictating every move to minimize exposure time. Shielding, time management, and access control are the stars of this performance, ensuring that the dance of atoms remains a safe and controlled spectacle.

Act III: The Radioactive Retreat

Our final act takes us to the resting place of radioactive materials – storage facilities. These are the fortresses of nuclear security, where shielded vaults and bunkers stand guard against accidental exposure and environmental contamination. Inventory control and waste management are the diligent caretakers, ensuring that these materials remain safely confined until their energy fades away.

Epilogue: The ALARA Principle

Throughout our journey, one principle has been our guiding star – the ALARA principle, which stands for "As Low as Reasonably Achievable." This principle is the mantra of radiation safety, reminding us to minimize exposure time and keep radiation doses as low as possible.

Technological advancements are also joining the cast, with robotic systems and new imaging techniques taking the stage to further reduce human exposure. These innovations are the rising stars of nuclear safety, promising a future where the benefits of radioactive materials can be enjoyed with even greater confidence.

Curtain Call

As our radioactive rendezvous comes to an end, we're left with a profound appreciation for the delicate balance between harnessing the power of radioactive materials and ensuring the safety of those who handle them. The journey has been a testament to human ingenuity and our unwavering commitment to safety in the face of invisible forces. So, let's give a round of applause to the unsung heroes of nuclear safety, who work tirelessly to ensure that the atomic dance remains a safe and captivating performance.

Keeping Our Distance: A Radioactive Safety Dance

Imagine radioactive materials as tiny, energetic dancers, constantly throwing out invisible sparks. These sparks, or radiation, can be helpful in small doses (think medical scans or x-rays), but too much can be harmful. That's where the "radioactive safety dance" comes in! The most important step? Distance, distance, distance!

Think of it like a campfire. Get too close, and you'll get burned. Back away, and you're safe and sound, enjoying the warmth. The same goes for radiation. The further you are from the source, the lower your exposure. It's all thanks to the inverse square law – a fancy way of saying that doubling your distance cuts the radiation by a whopping 75%!

Moving those Energetic Dancers:

Transporting radioactive materials is like a delicate ballet. Every step is carefully choreographed to keep everyone safe.

Special Packaging: Imagine these dancers are packed in super-strong, shielded boxes. Some boxes are like cozy studios (Type A) for less energetic dancers, while others are like fortified theaters (Type B) for the real showstoppers.
Strategic Routes: No crowded parades here! We plan routes that are like quiet, secluded pathways, minimizing the number of people nearby.
Separate Stages: Just like rival dance troupes, different radioactive materials are kept apart to avoid their "sparks" combining into a bigger show.

Putting those Dancers to Work:

Radioactive materials are versatile performers, taking the stage in medicine, industry, and research.

Medical Marvels: In hospitals, these dancers help doctors see inside the body (think PET scans) or even fight off diseases like cancer (with radiotherapy). It's like a precisely choreographed medical performance, with minimal exposure for patients and staff.
Industrial Ingenuity: In factories, these dancers inspect welds, measure materials, and even control processes. They're like invisible quality control experts, working behind the scenes with proper shielding to keep everyone safe.
Research Revelations: Scientists use tiny amounts of these dancers to unlock the secrets of the universe, from tracking how plants absorb nutrients to understanding the building blocks of matter. It's like a microscopic ballet, revealing the hidden wonders of nature.
Giving those Dancers a Rest:

When the performance is over, it's time for a well-deserved rest.

Secure Storage: Radioactive materials are stored in special facilities, like high-security dressing rooms, with thick walls and strict access control.
Organized Inventory: Knowing exactly where each dancer is and what kind of "spark" they emit helps minimize the time anyone spends near them.

Beyond the Distance:

While distance is the star of the show, other safety measures play supporting roles:

Time Limits: Like any good performance, limiting exposure time is key.
Protective Gear: Shielding and personal protective equipment are like costumes that provide an extra layer of protection.
Cleanliness: Preventing contamination is like keeping the stage clean and tidy, ensuring no stray "sparks" are left behind.

By mastering the "radioactive safety dance," we can harness the incredible power of these energetic dancers while keeping ourselves and the environment safe. It's a delicate balance, but with careful planning and continuous improvement, we can enjoy the show without getting burned.

Radioactive Adventures: Shielding Ourselves from Invisible Danger!

Imagine tiny ninjas, invisible to the naked eye, shooting out from glowing rocks! Some are slow and clumsy, easily stopped by a sheet of paper (like those embarrassing foam ninja stars). Others are faster, zipping through things like tiny bullets. And then there are the super-ninjas, the gamma rays, shooting through walls like they're not even there!

That's kind of what radioactive materials is like – they throw out these invisible ninjas, called radiation, that can be pretty harmful. That's where "shielding" comes in: it's like building a fortress to keep those pesky ninjas at bay!

1. Know Your Enemy: Radiation and its Quirks

Before we start building our fortress, we need to know what we're up against. These radiation ninjas come in different flavors:

Alpha particles: The slowpokes of the bunch. Imagine them tripping over their own feet! A sheet of paper is enough to stop them. But, if they sneak into your body (say, you breathe them in), they can cause some real trouble.
Beta particles: These guys are like tiny speed demons, zipping around! They can even go through aluminum foil! Definitely don't want to get hit by these.
Gamma rays: The ultimate ninjas! They can pass through walls like ghosts. We need some serious heavy-duty materials to stop these guys, like thick concrete or lead.
Neutron radiation: These are like the weird cousins of the radiation ninjas. They're rare, mostly hanging out in nuclear power plants, and they need special materials like water to stop them.

2. Building Our Fortress: Shielding Materials

Now, let's talk about building our fortress! We need strong, sturdy materials to block those radiation ninjas. Think of it like this:

Lead: The ultimate ninja-proof vest! Super dense and tough, it's like wearing a suit of armor against gamma rays. Doctors and scientists use it all the time.
Concrete: Like building a castle wall! It's thick, heavy, and great for stopping gamma rays and those weird neutron cousins.
Water: Surprisingly, water is like a ninja trap for neutrons! They get stuck in it and can't escape. That's why nuclear reactors are often submerged in water.
Steel: Strong and reliable, like a trusty shield. It's good for some protection, but we often combine it with other materials for extra strength.
Tungsten: Even denser than lead! This stuff is like super-armor, often used in hospitals for cancer treatments.
Depleted Uranium: The ultimate ninja-stopping material! It's so dense, it's like a black hole for radiation. We use it for transporting really dangerous radioactive stuff.

3. Radioactive Road Trip: Shielding on the Move!

Ever seen those trucks with the warning signs, carrying mysterious cargo? They're probably transporting radioactive materials! To keep those ninjas from escaping on the road, we use special containers:

Type A packages: Like a padded envelope for small, not-so-dangerous stuff.
Type B packages: Heavy-duty boxes for more dangerous materials. They can survive crashes and fires!
Type C packages: The ultimate armored trucks! Designed for the most dangerous stuff, like spent nuclear fuel. They could probably survive a meteor shower!

4. Using Radioactive Ninjas for Good: Shielding in Action

Believe it or not, we can actually use these radiation ninjas for good! But we need to be careful. Here's how we use shielding in everyday life:

Hospitals: X-ray machines, CT scanners, and cancer treatments all use radiation. Lead aprons and shields protect patients and doctors.
Factories: Radiation can be used to check for cracks in metal or measure the thickness of materials. Workers use shields and special equipment to stay safe.
Science labs: Scientists use radiation for all sorts of cool experiments! They work behind lead walls and use special boxes to keep the ninjas contained.

5. Locking Up the Ninjas: Shielding in Storage

When we're not using those radioactive ninjas, we need to keep them locked up tight!

Nuclear power plants: Spent nuclear fuel is stored in giant pools of water or super-strong casks to keep it safe.
Hospitals: Radioactive materials are kept in lead-lined vaults, like something out of a spy movie!
Labs: Scientists store their radioactive materials in special containers, like tiny prisons for ninjas.

6. Oops! When Things Go Wrong...

Sometimes, even with the best fortresses and shields, accidents happen. Remember the Fukushima disaster in Japan? The tsunami flooded a nuclear power plant, and those radiation ninjas escaped! It was a reminder that we always need to be careful and improve our defenses.

So, there you have it! A crash course in radioactive materials and the amazing world of shielding. It's all about keeping those invisible ninjas under control and using their power for good, while staying safe!

Space physics

Earth and the Solar System

The Earth

Earth's Dizzying Dance: A Whirlwind Tour Around Our Star

Hold on tight! We're about to embark on a cosmic joyride, exploring Earth's wild and wonderful journey around the Sun. Forget boring textbooks and dry lectures – this is a vibrant voyage through space, where we'll uncover the secrets of our planet's orbital antics.

Forget Circles, Think Ovals!

First things first, let's ditch the outdated idea of a perfectly circular orbit. Earth's path is more like an oval racetrack, with the Sun acting as a radiant trophy at one end. This means our distance from the Sun is constantly changing, like a cosmic game of hot-and-cold. We're closest in January (perihelion), feeling the Sun's warm embrace, and farthest in July (aphelion), enjoying a bit of summer coolness.

Speeding Up, Slowing Down, never a Dull Moment!

Imagine Earth as a cosmic dancer, twirling around the Sun with ever-changing speed. Like a figure skater pulling their arms in for a faster spin, Earth speeds up when it's closer to the Sun and slows down when it's farther away. It's a delicate balancing act, ensuring we don't get sucked into the Sun's fiery embrace or flung out into the cold depths of space.

The Ecliptic: Earth's Celestial Highway

Earth's orbital path isn't just a random route through space; it follows a specific plane called the ecliptic. Think of it as Earth's own personal highway, guiding its journey around the Sun. And guess what? Our planetary neighbors are cruising along on similar paths, creating a cosmic freeway of sorts.

Tilted and Terrific: The Reason for Seasons

As Earth zips around the Sun, it's also spinning like a top, tilted at a jaunty angle. This tilt is the secret ingredient behind our beloved seasons. It's like Earth is showing off different sides to the Sun throughout the year, resulting in the warm, sunny days of summer and the cozy, snowy days of winter.

A Year in the Making: Earth's Orbital Odyssey

One complete loop around the Sun marks a year on Earth, but there are different ways to measure this cosmic lap. The sidereal year is like timing Earth against the backdrop of distant stars, while the tropical year is all about the cycle of seasons. And then there's the calendar year, our trusty guide for birthdays and holidays, with a little leap day thrown in to keep things accurate.

Cosmic Tug-of-War: Influences on Earth's Orbit

Earth's orbit isn't a solo performance; it's a cosmic dance influenced by the gravitational pull of other celestial bodies. The Sun and Moon are the main choreographers, but even distant planets have a say, causing subtle shifts in Earth's path over vast stretches of time.

Climate Change and Earth's Wobbly Path

Earth's orbit isn't static; it undergoes slow, rhythmic changes known as Milankovitch cycles. These cycles, driven by the gravitational pull of other planets, can trigger dramatic shifts in Earth's climate, including ice ages. It's a reminder that our planet's journey is a dynamic and ever-changing one, with profound implications for life on Earth.

Earth's Orbital Adventure: A Never-Ending Story

Earth's orbit is a captivating tale of cosmic motion, shaped by gravity, momentum, and the interplay of celestial bodies. It's a journey that dictates our seasons, influences our climate, and connects us to the vastness of the universe. So, buckle up and enjoy the ride as Earth continues its dizzying dance around the Sun, a testament to the wonders of our solar system.

Our Wacky, Wobbly Ride Around the Sun

Forget what they told you in school! Earth's orbit isn't some boring, perfect circle. Nope, it's more like a slightly squashed oval, an ellipse, if you want to be fancy. And this cosmic quirkiness has some pretty wild effects on our planet.

Ellipses: Not Just for Breakfast Anymore

Imagine an oval. Got it? Now, picture two dots inside that oval. Those are the foci (focuses for the grammar sticklers). One of those dots is our very own Sun, keeping us warm from afar. The other? Well, let's just say it's an invisible friend, hanging out in space.

Now, the more stretched out that oval is, the crazier our orbit. Luckily for us, Earth's orbit is pretty chill, almost a circle. But that "almost" makes all the difference!

Cosmic Tag with the Sun

Remember playing tag as a kid? Well, Earth is playing a cosmic game of tag with the Sun. Sometimes we're closer (perihelion - say that five times fast!), around January 3rd, like we're trying to give the Sun a high five. Other times, around July 4th, we're hanging back (aphelion), maybe grabbing a cosmic popsicle to cool down.

But here's the thing: Earth doesn't just stroll around the Sun. It speeds up and slows down like a kid on a rollercoaster! The closer we are, the faster we zoom. It's like the Sun's gravity is giving us an extra push.

Why Should You Care?

So, why does this matter to you, sitting there reading this? Well, this elliptical dance affects everything from the seasons (though Earth's tilt is the real MVP there) to the long-term fate of our planet.

Think ice ages! Over thousands of years, our orbit gets even wonkier thanks to the gravitational pull of the other planets, like those cosmic bullies Jupiter and Saturn. This can trigger massive climate shifts, turning Earth into a giant snowball. Brr!

Keeping an Eye on Our Cosmic Dance

Thankfully, scientists are like cosmic detectives, using all sorts of high-tech tools to track our orbit. They're keeping tabs on our wild ride, making sure we don't go careening off into space.

So next time you look up at the Sun, remember: it's not just a static ball of fire. We're in a dynamic, ever-changing relationship with our star, and that's something to marvel at. It's a cosmic dance, a wobbly, wacky journey that makes life on Earth the exciting adventure it is.

The Earth's Spin: A Cosmic Ballet

Imagine our planet as a ballerina, gracefully twirling in the vastness of space. This twirl, known as Earth's rotation, is a mesmerizing dance that dictates the rhythm of our lives. It's the reason we wake to the sun's warm embrace and slumber under a blanket of stars.

The Earth's Pirouette: A Closer Look

Our ballerina spins on an invisible axis, a line connecting the North and South Poles. This pirouette takes roughly 24 hours to complete, creating the day and night cycle we all know. But there's more to this dance than meets the eye.

A Tilted Stage: Earth's axis isn't perfectly upright. It's tilted at a jaunty angle of 23.5 degrees, like a dancer leaning into a graceful pose. This tilt is what gives us our seasons, as different parts of the planet bask in varying amounts of sunlight throughout the year.

A Speedy Waltz: While we perceive a 24-hour rotation, Earth is actually a bit quicker on its feet. It completes a rotation relative to the stars in about 23 hours, 56 minutes, and 4 seconds. This slight discrepancy arises from Earth's simultaneous journey around the sun.

An Unpredictable Rhythm: Like any good dancer, Earth isn't perfectly consistent. The gravitational pull of the moon and sun, along with internal rumblings and even weather patterns, can cause slight variations in its spin. These variations are tiny but measurable, adding a touch of improvisation to the cosmic ballet.

The Impact of the Dance

Earth's rotation isn't just a beautiful spectacle; it's a driving force that shapes our world:

Day and Night: The most obvious consequence of this celestial dance is the cycle of day and night. As Earth twirls, different parts of the planet face the sun, experiencing daylight, while others are shrouded in darkness. This rhythm is essential for life, regulating everything from our sleep patterns to the behavior of plants and animals.

The Coriolis Effect: Imagine a dancer spinning with outstretched arms. As they spin, objects they hold will appear to curve. This is similar to the Coriolis effect, where moving objects on Earth curve due to the planet's rotation. This effect influences weather patterns, ocean currents, and even the flight paths of airplanes.

Tides: The moon's gravitational pull tugs at Earth's oceans, creating bulges of water that we experience as tides. Earth's rotation carries different regions through these bulges, influencing the timing and location of tides.

Earth's Shape: The centrifugal force of Earth's spin causes it to bulge slightly at the equator and flatten at the poles, giving it a shape known as an oblate spheroid. It's like a dancer's skirt flaring out as they spin.

Magnetic Field: Earth's rotation plays a crucial role in generating our planet's magnetic field, a protective shield against harmful solar radiation. It's like a dancer's energy creating an invisible aura around them.

Measuring the Dance

Scientists use various techniques to monitor Earth's graceful performance:

Very Long Baseline Interferometry (VLBI): This technique uses radio telescopes to observe distant quasars, providing precise measurements of Earth's orientation and rotation.

Satellite Laser Ranging (SLR): SLR involves bouncing lasers off satellites to track their positions and infer information about Earth's rotation.

Global Navigation Satellite Systems (GNSS): GNSS, including GPS, can be used to monitor Earth's rotation by analyzing signals from multiple satellites.

Examples of the Dance in Action

Foucault Pendulum: This simple experiment demonstrates Earth's rotation. The pendulum swings back and forth, but as Earth rotates beneath it, the plane of oscillation appears to rotate.

Hurricane Formation: The Coriolis effect plays a crucial role in the formation of hurricanes, causing the swirling air masses to organize into powerful storms.

Ocean Gyres: Large systems of rotating ocean currents, known as gyres, are influenced by the Coriolis effect, creating vast whirlpools in the oceans.

Leap Seconds: Due to slight variations in Earth's rotation, leap seconds are occasionally added to Coordinated Universal Time (UTC) to keep our clocks in sync with the planet's rhythm.

The Final Bow

Earth's rotation is a dynamic and complex dance that shapes our world in countless ways. It's a reminder of our planet's constant motion and the delicate balance that sustains life. By understanding this cosmic ballet, we gain a deeper appreciation for the intricate workings of our home in the universe.

The Moon's Dance: A Celestial Ballet with Earth

Imagine a cosmic dance floor where Earth takes center stage, bathed in the warm glow of the Sun. Enter the Moon, our celestial partner, twirling and swirling in a captivating waltz around our planet. This isn't just any ordinary dance; it's a performance billions of years in the making, a ballet of gravity and motion that shapes our world in countless ways.

A Month of Many Faces

The Moon's journey around Earth is a story told in chapters, each a different kind of "month." There's the sidereal month, a swift 27.32-day sprint where the Moon races against the backdrop of distant stars. Then there's the synodic month, a more leisurely 29.53-day stroll that marks the familiar cycle of lunar phases, from the shy new moon to the radiant full moon and back again.

But the Moon's dance isn't always predictable. The anomalistic month sees our lunar companion drawing closer, then retreating, like a lover testing the boundaries of intimacy. And the draconic month reveals a hidden pattern, a weaving path where the Moon's orbit intersects with Earth's own journey around the Sun, setting the stage for dramatic eclipses.

The Pull of Partners and the Sway of the Sun

The Moon's graceful movements are guided by an invisible hand – the force of Earth's gravity. It's a powerful embrace that keeps the Moon from spinning off into the vastness of space. But there's another player in this celestial dance: the Sun. Its gravity tugs at the Moon, adding a subtle sway to its steps and shaping the intricate patterns of its orbit.

The Moon's Influence: More Than Just Moonlight

This celestial ballet isn't just a spectacle for stargazers; it's a performance that touches our lives in profound ways. The Moon's gravity, like an invisible conductor, orchestrates the ocean tides, drawing water towards it in a rhythmic rise and fall. It also acts as a stabilizing force, keeping Earth's axial tilt in check and ensuring a more stable climate.

And let's not forget the magic of eclipses! When the Sun, Earth, and Moon align in perfect harmony, they create breathtaking displays of shadow and light, reminding us of the intricate connections that bind our universe together.

Exploring the Moon's Rhythms

Scientists have become keen choreographers of this celestial dance, using tools like lunar laser ranging to track the Moon's every move with incredible precision. They've uncovered secrets hidden in the Moon's rhythmic journey, like the Saros cycle, a repeating pattern of eclipses that unfolds over nearly two decades.

A Timeless Dance

The Moon's dance with Earth is a timeless story, a testament to the enduring power of gravity and the beauty of celestial motion. It's a story that continues to unfold above us every night, inviting us to look up and marvel at the wonders of our universe.

Imagine a Sunbeam's Wild Ride to Earth!

Ever basked in the warm glow of the sun and wondered about the incredible journey that sunlight takes to reach our little planet? Buckle up, because we're about to dive headfirst into a cosmic adventure that's more thrilling than a rollercoaster ride!

Light Speed Ahead!

First things first, let's talk speed. Light doesn't just travel fast; it travels at the ultimate speed limit of the universe – a mind-boggling 299,792,458 meters per second! To put that into perspective, it could zoom around Earth's equator over seven times in a single second. Zoom!

The 8-Minute Time Machine

Now, here's the kicker: even at that breakneck speed, it still takes sunlight about 500 seconds – or roughly 8 minutes and 20 seconds – to reach us. That means the sunlight warming your face right now actually left the sun's surface over 8 minutes ago! We're literally basking in the past, dude!

A Photon's Epic Quest

Let's follow the journey of a single photon, a tiny particle of light, as it embarks on its epic quest from the sun's fiery heart to your sun-kissed skin:

Born in a Nuclear Furnace: Our photon is born in the sun's core, a chaotic mosh pit of atoms where hydrogen nuclei are fused together to form helium, unleashing a torrent of energy in the process.

The Radiative Zone Rumble: Next, our photon braves the radiative zone, a dense, swirling mass of plasma where it's constantly jostled and bumped around like a tiny ball in a pinball machine. This crazy dance can last for hundreds of thousands, even millions of years!

Convective Chaos: Finally, our photon reaches the convective zone, where it hitches a ride on a giant plasma current, like a surfer catching a gnarly wave.

Escape from the Photosphere: With a burst of energy, our photon breaks free from the sun's surface – the photosphere – and launches itself into the vast expanse of space. Freedom!

The Cosmic Cruise: Our photon cruises through the vacuum of space at light speed, dodging asteroids, comets, and rogue planets like a cosmic ninja.

Earth's Atmospheric Obstacle Course: As it approaches Earth, our photon encounters the atmosphere, a swirling cocktail of gases, water vapor, and dust. It bounces and scatters, painting the sky with vibrant hues of blue and fiery sunsets.

Touchdown! Finally, our photon reaches its destination – your skin! It's absorbed, its energy transformed into warmth, giving you that sun-kissed glow.

Cosmic Consequences and Mind-Bending Implications

This 8-minute delay isn't just a fun fact; it has some serious implications:

Space Weather Watch: When the sun throws a tantrum – like a solar flare or coronal mass ejection – we have a little over 8 minutes to brace ourselves for impact. This warning time is crucial for protecting our satellites and power grids from potential damage.

Peering into the Past: Astronomers have to account for this time lag when observing solar events. They're essentially looking back in time, like cosmic historians!

Interstellar Chit-Chat: If we ever want to chat with aliens, we'll need to be patient. Messages will take years, decades, or even centuries to reach them, depending on how far away they are. Talk about a long-distance call!

Beyond Our Solar System: Light-Years and Cosmic Distances

The 8-minute journey from the sun to Earth is a mere hop, skip, and a jump compared to the vast distances between stars and galaxies. To measure these cosmic expanses, we use light-years – the distance light travels in a year. That's a whopping 9.46 trillion kilometers!

The closest star system to ours, Alpha Centauri, is over 4 light-years away. So, the light we see from it today actually left those stars over 4 years ago! Mind-blowing, right?

The Universe in a Grain of Sunlight

So, the next time you feel the sun's warmth on your skin, remember the incredible journey that sunlight takes to reach you. It's a story of cosmic proportions, filled with adventure, danger, and mind-boggling distances. And it all happens in the blink of an eye – or, more accurately, in 8 minutes and 20 seconds!

Cosmic Waltz: Unveiling the Secrets of Orbital Speed

Imagine the universe as a grand ballroom, with celestial bodies twirling and swirling in an intricate cosmic waltz. Each object, from the smallest asteroid to the largest galaxy, is engaged in a perpetual dance, gracefully orbiting around a central partner. The speed at which these objects traverse their orbital paths is a fundamental concept in astrophysics, providing a glimpse into the invisible forces that orchestrate this celestial ballet.

Orbital Speed: The Tempo of the Cosmic Dance

Just as the tempo of a waltz dictates the speed of the dancers, the orbital speed of a celestial body determines how quickly it moves along its orbital path. This speed is not arbitrary; it's a delicate balance between the object's inertia, its desire to move in a straight line, and the gravitational pull of its central partner, which constantly tugs it inward.

The average orbital speed is like the average tempo of a waltz, taking into account the overall speed of the dance. It's calculated by dividing the total distance traveled by the object along its orbit by the time it takes to complete one full revolution, known as the orbital period.

The Equation: A Mathematical Pas de Deux

The equation for average orbital speed is a simple yet elegant formula that encapsulates the relationship between the object's speed, the radius of its orbit, and its orbital period:

$$v = 2\pi r / T$$

where:

v is the average orbital speed, the swiftness of the celestial dancer.
r is the average radius of the orbit, the distance between the partners in this cosmic dance.
T is the orbital period, the time it takes for one complete twirl around the dance floor.

This equation reveals that the farther apart the partners are, the slower their dance, and the longer it takes to complete a revolution. Conversely, the closer they are, the faster their dance, and the shorter the orbital period.

Why Orbital Speed Matters

Understanding orbital speed is like understanding the language of the cosmic waltz. It allows us to decipher the gravitational forces at play, the energy of the dancers, and the overall stability of their intricate movements.

By analyzing the orbital speed of planets, moons, and stars, scientists can unravel the mysteries of planetary systems, the distribution of mass within galaxies, and even the presence of exoplanets – planets that orbit distant stars, hidden from our direct view.

Factors Influencing the Cosmic Dance

Just as the tempo of a waltz can be influenced by the music, the dancers' skill, and the size of the dance floor, several factors can affect the orbital speed of a celestial body:

Mass of the Central Body: The more massive the central partner, the stronger its gravitational pull, and the faster the orbiting object needs to move to avoid being drawn in.
Radius of the Orbit: As we've seen, the farther apart the partners, the slower their dance.
Shape of the Orbit: While the equation for average orbital speed assumes a circular orbit, most celestial dances are more elliptical, with the speed varying throughout the orbit.

Examples: A Glimpse into the Cosmic Ballroom

Earth's Orbit: Our planet waltzes around the Sun at an average speed of 29.78 km/s, completing one full revolution in approximately 365.25 days.
Geostationary Satellites: These satellites perform a unique dance, orbiting Earth at a specific altitude and speed that allows them to

remain stationary relative to a point on the ground, making them ideal for communication and navigation.
Exoplanet Detection: By observing the subtle wobbles of distant stars, astronomers can infer the presence of exoplanets and even estimate their orbital parameters, including their average orbital speed.

Conclusion: The Dance Goes On

The concept of average orbital speed is a fundamental key to understanding the intricate dynamics of the universe. It allows us to appreciate the delicate balance between inertia and gravity, the interplay of mass and distance, and the elegant choreography of the cosmic waltz. As we continue to explore the vast expanse of space, the secrets of orbital speed will undoubtedly lead us to new discoveries and a deeper understanding of our place in the grand cosmic ballroom.

The Solar System

The Sun: Our Very Own Star

Imagine a gigantic, blazing ball of fire, so massive that it holds almost everything in our solar system in its grasp. That's our Sun! It's not just a pretty light in the sky; it's a churning, bubbling, exploding powerhouse that provides the energy for life on Earth.

A Closer Look

Even though we call it a "yellow dwarf," the Sun's light is actually pure white. It's only our atmosphere that makes it look yellow. And don't let the "dwarf" part fool you - this star is enormous! If the Sun were a hollow ball, you could fit over a million Earths inside.

Inside the Sun

The Sun is like an onion with different layers. At its heart is the core, where the real action happens. Imagine tiny atoms smashing together and creating a massive explosion of energy. That's nuclear fusion, and it's what makes the Sun shine. This energy then travels outwards through the radiative and convective zones, like a hot air balloon rising through the sky.

The Sun's Atmosphere

The Sun also has an atmosphere, but you wouldn't want to breathe it! The photosphere is the visible surface, and it's where we see sunspots, like freckles on the Sun's face. Above that is the chromosphere, a reddish layer where solar flares erupt, like sudden bursts of fireworks. And finally, there's the corona, a super-hot halo that you can only see during a total solar eclipse.

The Sun's Mood Swings

The Sun has a bit of a temper. Its magnetic field is constantly changing, causing sunspots, solar flares, and coronal mass ejections (CMEs). These are like the Sun's way of throwing a tantrum, and they can sometimes cause problems for us on Earth, like disrupting our radio signals or even causing power outages.

The Sun's Influence

The Sun is like the conductor of an orchestra, keeping all the planets, asteroids, and comets in our solar system moving in harmony. It also provides the energy for our weather and climate. And when the solar wind, a stream of charged particles from the Sun, interacts with Earth's magnetic field, it creates the beautiful auroras, or northern and southern lights.

The Sun's Life Story

Our Sun is about 4.6 billion years old, which is about middle-aged for a star. It's expected to keep shining for another 5 billion years before it starts to run out of fuel and eventually becomes a white dwarf, a small and faint ember in the vastness of space.

Exploring the Sun

Scientists are constantly trying to learn more about the Sun, sending spacecraft like the Parker Solar Probe and the Solar Orbiter to study it up close. It's a dangerous mission, but it's helping us understand our own star and its place in the universe.

The Sun: A Source of Wonder

The Sun is more than just a giant ball of gas; it's the lifeblood of our solar system. It provides the energy for life, drives our weather, and creates beautiful displays of light in our sky. By studying the Sun, we can learn more about our own planet and our place in the cosmos.

Our Stellar Neighborhood: A Quirky Guide to the Planets

Forget those boring old textbooks! We're diving deep into the solar system with a fresh perspective, exploring the personalities and quirks of our planetary neighbors.

1. Mercury: The Sun's Speedy Messenger

This little guy is the ultimate daredevil, zipping around the Sun faster than any other planet. It's a land of extremes, where scorching days meet frigid nights. Imagine a place where you could fry an egg on a rock in the morning and then need a parka to survive the evening chill!

2. Venus: The Sultry Siren

Don't let her beauty fool you! Venus is a scorching hot temptress with a toxic atmosphere that could melt lead. She's shrouded in a thick veil of clouds, hiding a landscape of volcanoes and lava flows. This is one planet you definitely wouldn't want to visit without a serious heat shield.

3. Earth: Our Home Sweet Home

Ah, Earth! The only planet we know of that's bursting with life. From the lush rainforests to the icy polar caps, our home planet is a vibrant tapestry of ecosystems and creatures. Let's not forget those breathtaking sunsets and the magical dance of the Northern Lights!

4. Mars: The Rusty Enigma

This red-hued planet has captivated our imaginations for centuries. Was it once a watery world teeming with life? The search for answers continues as rovers explore its ancient riverbeds and canyons. Maybe one day, humans will walk on Mars and uncover its secrets.

5. Jupiter: The King of the Planets

Jupiter is a colossal giant, a swirling vortex of storms and colorful clouds. Its iconic Great Red Spot is a storm so large it could swallow Earth whole! This gas giant is a true wonder of the solar system, with a retinue of moons that could rival a small solar system of their own.

6. Saturn: The Cosmic Jewel

Prepare to be dazzled by Saturn's breathtaking rings, a celestial ballet of ice and rock. This gas giant is a true showstopper, a reminder of the universe's infinite beauty. Imagine soaring through those rings, witnessing their intricate patterns and shimmering colors up close.

7. Uranus: The Sideways Wonder

Uranus is the quirky one, tilted on its side like a spinning top that's lost its balance. This ice giant is shrouded in a hazy atmosphere, hiding a frigid world of methane ice. It's a planet of mysteries, waiting to be explored.

8. Neptune: The Windy Wonderland

Neptune is the farthest planet from the Sun, a realm of icy winds and swirling storms. Its deep blue atmosphere is reminiscent of a stormy ocean, a fitting image for this distant and mysterious world.

Beyond the Known: The Solar System's Final Frontier

Our journey doesn't end with the eight planets. Beyond Neptune lies a realm of icy bodies and dwarf planets, including the enigmatic Pluto. And even farther out, the Oort Cloud, a vast sphere of comets, marks the outer limits of the Sun's influence.

Our Cosmic Home: A Never-Ending Story

The solar system is a place of endless wonder, a cosmic playground filled with mysteries and marvels. As we continue to explore its depths, we'll undoubtedly uncover even more secrets and surprises. So, keep your eyes on the stars, and let your imagination soar!

Cosmic Crumbs: A Whimsical Journey Through the Solar System's Leftovers

Our solar system isn't just a stately dance of eight planets waltzing around the sun. It's more like a cosmic party with a guest list bursting at the seams! Among the VIPs (Very Important Planets) mingle a motley crew of characters – the minor planets. Think of them as the quirky cousins, the unexpected plus-ones, the life of the interplanetary gathering.

What Makes a planet "Minor"?

Imagine a planet as a cosmic vacuum cleaner, diligently tidying up its orbit by swallowing or flinging away any space debris. Minor planets, however, are a bit messier. They haven't quite mastered the art of cosmic housekeeping and share their orbital paths with a whole bunch of other objects.

A Menagerie of Minor Planets

1. Dwarf Planets: The Almost-Planets

These celestial underdogs are like planets who just missed the cut. They're round enough to be considered cool, but they haven't quite cleaned up their act in the orbital department.

Pluto: The most famous of the bunch, Pluto was demoted from planet status in 2006, causing quite a stir in the cosmic community.
Ceres: Hanging out in the asteroid belt, Ceres is a bit of a show-off, being the first dwarf planet to receive a spacecraft visit.
Eris: This troublemaker, slightly larger than Pluto, lurks in the scattered disk, a region beyond the Kuiper Belt, like a rebellious teenager hanging out at the edge of town.

Haumea: This oddball spins so fast that it's stretched out into an elongated shape. Talk about a cosmic workout!
Makemake: This icy loner resides in the Kuiper Belt and is known for its frosty demeanor and lack of any significant atmosphere.

2. Asteroids: Cosmic Leftovers

Asteroids are like the crumbs left over from the solar system's formation – rocky remnants from a bygone era. Most of them hang out in the asteroid belt, a busy highway between Mars and Jupiter.

Types of Asteroids:

C-type: The most common type, as dark and mysterious as a cosmic charcoal briquette.
S-type: Made of silicate materials, these asteroids sparkle like celestial gemstones.
M-type: Metallic asteroids, potentially hiding veins of iron and nickel, like cosmic treasure chests waiting to be opened.

Asteroid Celebrities:

Vesta: The second-largest asteroid, visited by the Dawn spacecraft, which probably snapped a few selfies while it was there.
Pallas: Known for its tilted orbit, Pallas is the cosmic equivalent of a skateboarder doing an ollie.
Hygiea: This large asteroid is so round that it might even be a dwarf planet in disguise.

3. Trans-Neptunian Objects (TNOs): The Outer Limits

These icy bodies are the ultimate cosmic hermits, dwelling in the frigid outskirts of the solar system, beyond the realm of Neptune.

Kuiper Belt Objects (KBOs): Imagine a donut-shaped realm filled with icy bodies, including Pluto, Haumea, and Makemake. That's the Kuiper Belt.
Scattered Disk Objects (SDOs): These rebels have wild, unpredictable orbits, like cosmic bumper cars bouncing around the solar system.

Oort Cloud Objects: These mysterious objects are thought to be the source of long-period comets, like cosmic snowballs flung from the farthest reaches of the solar system.

The Asteroid Belt: A Cosmic Demolition Derby

The asteroid belt is like a cosmic demolition derby, a chaotic jumble of asteroids zipping around between Mars and Jupiter. It's believed to be the remnants of a failed planet, a victim of Jupiter's massive gravitational pull.

Kirkwood Gaps: These are empty zones within the asteroid belt, like cosmic potholes, where Jupiter's gravity has cleared out the asteroids.
Trojan Asteroids: These asteroids share Jupiter's orbit, like loyal followers, hanging out at Lagrange points, where the gravitational forces are perfectly balanced.

The Kuiper Belt and Beyond: The Final Frontier

The Kuiper Belt is a vast, icy realm beyond Neptune, home to dwarf planets and countless KBOs. It's like the asteroid belt's bigger, colder sibling.

Resonant KBOs: These objects have orbits that are in sync with Neptune's orbit, like synchronized swimmers performing a cosmic ballet.
Scattered Disk: This region beyond the Kuiper Belt is a chaotic zone of eccentric orbits, like a cosmic mosh pit.
Oort Cloud: This hypothetical sphere of icy bodies surrounds the solar system at an incredible distance, like a cosmic shell encasing our little corner of the universe.

Why Should We Care About Minor Planets?

These cosmic crumbs might seem insignificant, but they hold valuable clues to the solar system's past, present, and future.

Cosmic Time Capsules: Their composition reveals secrets about the early solar system, like cosmic fossils preserving the past.

Space Rocks of Doom: Near-Earth asteroids could pose a threat to our planet, like cosmic missiles hurtling towards Earth.
Cosmic Treasure Troves: Asteroids might contain valuable resources like water and metals, potentially fueling future space exploration.
Windows to the Universe: Missions like Dawn, New Horizons, and OSIRIS have provided incredible insights into these fascinating objects, expanding our understanding of the cosmos.

Cosmic Case Studies

Ceres: The Dawn mission discovered evidence of past hydrothermal activity and organic molecules on Ceres, hinting at the possibility of ancient life.
Pluto: The New Horizons flyby revealed a surprisingly complex and diverse world with towering mountains, flowing glaciers, and a thin atmosphere.
Bennu: The OSIRIS mission collected samples from asteroid Bennu, which will be returned to Earth for analysis, potentially unlocking secrets about the origins of life.

The Future of Minor Planet Exploration

Exciting missions are planned to further explore these cosmic crumbs, including:

Lucy: This mission will visit multiple Trojan asteroids, like a cosmic archaeologist digging into the solar system's past.
Psyche: This mission will explore a metallic asteroid thought to be the exposed core of a protoplanet, like a cosmic geologist cracking open a planetary egg.
Europa Clipper: While focused on Jupiter's moon Europa, this mission will also gather data on Jupiter's Trojan asteroids, like a cosmic tourist taking a side trip to explore these intriguing objects.

In Conclusion

Minor planets may be small, but they hold immense significance in our understanding of the solar system. They are remnants of the past, potential sources of resources, and potential hazards. Continued exploration of these fascinating objects will undoubtedly reveal more secrets about our cosmic neighborhood, enriching our understanding of the universe and our place within it.

Moons: Our Solar System's Secret Worlds Unveiled

Forget the planets for a moment – our solar system is teeming with hidden gems, each a world of its own. These are the moons, the silent partners in a cosmic dance around our sun. They're more than just celestial sidekicks; they're a diverse cast of characters, each with its own story to tell.

A Lunar Family Reunion

Earth may be a "one-moon wonder," but our solar system is a moon-lover's paradise. Jupiter and Saturn, the gas giants, are practically drowning in moons, with new ones popping up faster than astronomers can name them. It's like a never-ending family reunion, and we're still meeting distant cousins!

Born from Chaos, Sculpted by Time

Moons are born from the same swirling chaos that birthed the planets themselves. Some coalesced from the leftover dust and gas, like cosmic leftovers transformed into gourmet meals. Others were cosmic wanderers, captured by a planet's gravity and forced into a lifelong waltz. And some? Well, they're the result of epic collisions, celestial fender-benders that left behind shimmering scars.

Meet the Lunar All-Stars

Earth's Moon: Our trusty companion, the ultimate mood setter. It controls our tides, stabilizes our wobble, and provides a romantic backdrop for countless love stories. Plus, it's the only place beyond Earth where humans have left their footprints.

Europa (Jupiter): This icy moon is hiding a secret – a vast ocean beneath its frozen shell. Imagine a world of underwater volcanoes, alien life forms swimming in the dark, and maybe even a hidden civilization of super-intelligent cephalopods. It's a sci-fi dream come true!

Titan (Saturn): Forget Mars, Titan is the real "second Earth." It has a thick atmosphere, rivers, lakes, and even rain. But hold on – the rivers are filled with methane, the lakes are ethane, and the rain is… well, let's just say you wouldn't want to get caught in it without a hazmat suit.

Ganymede (Jupiter): The biggest moon in the solar system, Ganymede is a planet in its own right. It's even got its own magnetic field, like a mini-Earth flexing its cosmic muscles.

Io (Jupiter): Io is the solar system's bad boy, a volcanic world covered in lava flows and sulfurous plumes. It's like a cosmic teenager, constantly erupting and throwing tantrums.

Enceladus (Saturn): This icy moon is like a giant soda bottle, spewing geysers of water vapor and ice particles into space. It's also got a hidden ocean, making it another prime candidate for alien life.

Triton (Neptune): Triton is the rebel of the moon family, orbiting Neptune in the opposite direction of its rotation. It's like a cosmic salmon swimming upstream, and its icy surface is dotted with nitrogen geysers.

Why Should We Care About Moons?

Moons are more than just pretty faces in the night sky. They're time capsules, preserving clues about the early solar system. They're laboratories, where we can study the conditions that might lead to life. And they're potential homes, offering new frontiers for human exploration.

The Future is Lunar

With missions like Europa Clipper, Dragonfly, and JUICE on the horizon, we're on the verge of a new era of lunar exploration. Who knows what secrets these missions will uncover? Maybe we'll find evidence of alien life, or maybe we'll discover new worlds that are even more bizarre and wonderful than we ever imagined.

So, the next time you look up at the night sky, don't just focus on the stars. Take a moment to appreciate the moons, those silent sentinels of the solar system. They're waiting to tell us their stories, if only we're willing to listen.

Cosmic Ballet: A Whimsical Tour of Comets and Moons

Our solar system is a bustling space metropolis, teeming with celestial bodies of all shapes and sizes. While the planets and the Sun may be the A-list celebrities, a captivating supporting cast of smaller objects adds depth and intrigue to this cosmic drama. Let's embark on a whimsical journey to explore the enchanting world of comets and moons, the unsung heroes of our solar system.

Comets: The Ethereal Vagabonds

Imagine a cosmic snow globe, filled with glistening ice crystals, dust particles, and rocky debris. That's essentially a comet, a frozen relic from the dawn of our solar system. These celestial nomads hail from the distant realms of the Kuiper Belt and the Oort Cloud, venturing into the inner solar system on their eccentric orbits.

The Cometary Anatomy:

Nucleus: The heart of a comet, a solid core composed of ice, dust, and rock. Think of it as a cosmic snowball, a few kilometers in diameter.
Coma: As a comet approaches the Sun, its icy nucleus begins to sublimate, forming a hazy cloud of gas and dust called the coma. It's like a celestial halo, enveloping the nucleus in a luminous glow.
Tail: The most dramatic feature of a comet, its tail, is formed by the solar wind and radiation pressure. It can stretch for millions of kilometers, a cosmic banner trailing behind the comet. The ion tail, composed of ionized gas, points directly away from the Sun, while the dust tail, made of dust particles, curves gracefully.

Cometary Varieties:

Short-Period Comets: These frequent visitors to the inner solar system have orbital periods of less than 200 years. They originate from the Kuiper Belt, a region beyond Neptune.

Long-Period Comets: These rare guests from the distant Oort Cloud take more than 200 years to complete an orbit. Their appearances are unpredictable and often spectacular.

Cometary Celebrities:

Halley's Comet: The most famous of all comets, Halley's Comet graces our skies every 76 years. Its last visit was in 1986, and it's scheduled to return in 2061. Mark your calendars!
Comet Hale-Bopp: This celestial showstopper was visible to the naked eye for a record 18 months in 1996 and 1997. Its brilliance and longevity captivated skywatchers worldwide.
Comet 67P/ Gerasimenko: This unassuming comet became a celestial superstar when the Rosetta spacecraft landed a probe on its surface in 2014, providing unprecedented insights into these icy wanderers.

Why Comets Matter:

Cosmic Time Capsules: Comets are like time capsules, preserving the pristine conditions of the early solar system. Studying them helps us understand how our planetary neighborhood formed and evolved.
Life's Building Blocks: Some scientists believe that comets may have delivered water and organic molecules to the early Earth, potentially contributing to the origin of life.
Cosmic Hazards: While most comets pose no threat to Earth, some have orbits that intersect with our planet's path. A collision with a large comet could have devastating consequences.

Moons: The Planetary Entourage

Moons, or natural satellites, are the loyal companions of planets, dwarf planets, and even asteroids. They come in a dazzling array of sizes, shapes, and compositions, each with its own unique story to tell.

Lunar Lineages:

Regular Moons: These well-behaved moons formed from the same disk of material as their parent planet. They orbit in the same direction as the planet's rotation, like a graceful dance.

Irregular Moons: These rebellious moons were captured by their parent planet's gravity. They often have eccentric orbits and may even orbit in the opposite direction of the planet's rotation, adding a touch of chaos to the cosmic ballet.

Lunar Luminaries:

Earth's Moon: Our faithful companion, the Moon, plays a vital role in stabilizing Earth's climate and tides. It's also a source of wonder and inspiration, a celestial beacon in our night sky.
Jupiter's Galilean Moons: Io, Europa, Ganymede, and Callisto, these four giants of the Jovian system are a fascinating quartet. Io is a volcanic inferno, Europa harbors a subsurface ocean, Ganymede is the largest moon in the solar system, and Callisto is a heavily cratered relic.
Saturn's Titan: This enigmatic moon boasts a dense atmosphere and methane lakes, making it a prime candidate for the search for extraterrestrial life.
Neptune's Triton: A captured moon with a retrograde orbit, Triton is a world of active geysers and icy plains.

Why Moons Matter:

Planetary Architects: Studying moons helps us understand how planetary systems form and evolve. They are like cosmic blueprints, revealing the intricate processes that shaped our solar system.
Cosmic Habitats: Some moons, like Europa and Enceladus, have subsurface oceans that could potentially harbor life. These hidden realms offer tantalizing possibilities for the search for extraterrestrial life.
Exploration Frontiers: Moons are exciting destinations for future space exploration. They offer diverse environments to explore, from volcanic landscapes to icy plains, and may hold the key to unlocking the secrets of our cosmic origins.

Case Studies: Cosmic Close-Ups

The Rosetta Mission and Comet 67P/ Gerasimenko: This groundbreaking mission provided a close-up look at a comet, revealing

its composition, structure, and activity. It's like receiving a postcard from the early solar system.
The Juno Mission and Jupiter's Moons: This ongoing mission is providing stunning images and data on Jupiter and its moons, including the Galilean moons and the Great Red Spot. It's like having a front-row seat to the Jovian spectacle.
The Europa Clipper Mission: This future mission will explore Jupiter's moon Europa, searching for evidence of a subsurface ocean and its potential to support life. It's like embarking on a cosmic treasure hunt.

Conclusion: The Cosmic Dance Continues

Comets and moons, the unsung heroes of our solar system, are captivating objects that offer valuable insights into our cosmic origins and our place in the universe. They are like celestial dancers, twirling and swirling in a cosmic ballet, each with its own unique rhythm and story to tell. As we continue to explore the solar system, these smaller bodies will undoubtedly play a crucial role in unraveling the mysteries of our cosmic neighborhood and our place in the grand tapestry of the cosmos.

A Cosmic Ballet: Unveiling the Secrets of Planetary Personalities

Imagine the solar system as a grand cosmic ballet, with each planet a unique dancer, twirling and leaping in a mesmerizing performance orchestrated by gravity. Each planet has its own distinct personality, shaped by its individual characteristics. Let's take a closer look at these fascinating performers and learn their secrets.

1. The Dance of Distance and Time: Orbital Distance and Period

Like dancers gracefully circling the stage, planets waltz around their star in a delicate balance of distance and time. Their orbital distance, the space between them and their star, dictates the length of their orbital period, the time it takes to complete one full circle.

Imagine Mercury, the fleet-footed messenger, zipping around the Sun in a quick 88 Earth days. Contrast that with distant Neptune, the slow and graceful giant, taking a leisurely 165 years to complete one orbit.

This cosmic choreography is governed by Kepler's Third Law, a celestial rule that links a planet's orbital distance to its period. It's like a mathematical metronome, keeping the planets in sync with the music of the cosmos.

2. The Weight of Worlds: Density

Just as dancers come in different shapes and sizes, planets also vary in their density, a measure of how much mass is packed into their volume. This characteristic provides clues about their inner makeup, revealing whether they are rocky, gaseous, or a mix of both.

Earth, our home planet, is a dense, rocky world, like a ballerina with strong, muscular legs. In contrast, Saturn, the ringed beauty, is a gas giant with a density so low it could float on water, like a dancer with light, airy movements.

3. A Planetary Fever: Surface Temperature

Each planet has its own unique temperature, a measure of its surface heat. This characteristic is influenced by a variety of factors, including its distance from the star, the presence of an atmosphere, and its albedo, or reflectivity.

Venus, the scorching hot goddess, is shrouded in a thick atmosphere that traps heat, making it the hottest planet in our solar system. Meanwhile, Mars, the chilly red warrior, has a thin atmosphere and a low albedo, resulting in a frigid surface.

4. The Pull of Home: Gravitational Field Strength

Like a dancer's connection to the stage, a planet's gravitational field strength keeps everything grounded. This force, determined by the planet's mass and radius, dictates how strongly it attracts objects on its surface.

Jupiter, the mighty king of the planets, boasts the strongest gravitational pull in our solar system, like a dancer with an undeniable stage presence. In contrast, Mercury, the smallest planet, has a relatively weak gravitational field, like a dancer with delicate, feather-light steps.

The Intertwined Dance of Planetary Characteristics

These planetary characteristics are not isolated traits but rather intertwined aspects of a planet's personality. They influence each other in a complex dance, shaping the planet's overall character.

For instance, a planet's density affects its gravitational field strength, while its distance from the star influences its surface temperature. These interrelationships provide valuable insights into a planet's formation, evolution, and potential for harboring life.

Unveiling the Mysteries of the Cosmos

By studying these planetary personalities, we can unlock the secrets of our solar system and the countless planetary systems beyond. Each planet, with its unique characteristics and story, contributes to the grand cosmic ballet, a mesmerizing performance that continues to captivate and inspire us.

As we continue to explore the universe, we will undoubtedly encounter even more diverse and fascinating planetary personalities, each with its own tale to tell. This ongoing journey of discovery promises to unveil even more wonders of the cosmos, enriching our understanding of our place in the universe.

Imagine a cosmic dance floor where planets twirl and stars shimmer. Gravity is the music that guides their every move, an invisible force that draws them together in a graceful waltz.

Now, picture each planet as a giant magnet, its strength determined by its mass. The more massive the planet, the stronger its magnetic pull, and the more intensely it attracts its dance partners – moons, asteroids, even passing comets.

But there's a twist! The size of the dance floor also matters. If the planet is a sprawling ballroom, the force of attraction weakens with distance. It's like trying to hold hands with your partner from across a crowded room – the further apart you are, the less you feel the connection.

This cosmic dance is governed by Newton's Law of Universal Gravitation, a mathematical formula that describes the attraction between any two objects with mass. It's like a secret code that unlocks the mysteries of the universe, revealing how planets stay in orbit and apples fall from trees.

Think of Earth and Mars as two dancers with very different styles. Earth, the heavyweight, commands attention with its strong gravitational pull, keeping us firmly grounded and holding onto our precious atmosphere. Mars, lighter on its feet, exerts a weaker force, making it easier for its atmosphere to drift away into space.

But gravity isn't just about keeping our feet on the ground. It's a powerful force that shapes entire worlds. On the moon, where gravity is a mere whisper, astronauts can leap like gazelles and objects weigh a fraction of their earthly counterparts. On Jupiter, a gas giant with a gravitational grip 2.5 times stronger than Earth's, even the most agile dancer would feel weighed down.

And then there are the extreme dancers, the neutron stars. These incredibly dense remnants of massive stars pack a gravitational punch that can bend light itself, warping the fabric of spacetime like a cosmic funhouse mirror.

Gravity's influence extends far beyond individual planets. It's the driving force behind tides, the rhythmic rise and fall of ocean waters caused by the gravitational tug-of-war between Earth, the moon, and the sun. It's the invisible hand that guides planets in their orbits, keeping them from spinning off into the vastness of space.

To measure this invisible force, scientists use ingenious tools like pendulums, free-falling objects, and even specialized instruments called gravimeters. They've discovered that gravity isn't uniform, varying with altitude, latitude, and even the density of the Earth's crust. It's like a cosmic symphony, with subtle variations in tempo and rhythm across the dance floor.

But the true maestro of this cosmic ballet is Einstein's theory of General Relativity, which describes gravity not as a force but as a curvature of spacetime caused by mass and energy. Imagine a bowling ball placed on a trampoline, causing the fabric to sag. That's how massive objects warp the fabric of spacetime, dictating how objects move and interact.

The study of gravity is a journey into the heart of the universe, a quest to understand the fundamental forces that shape our existence. From the everyday experience of weight to the mind-bending concepts of black holes and the expanding universe, gravity is a constant reminder of the interconnectedness of all things. It's a cosmic dance that has been going on for billions of years, and it's a dance that will continue long after we're gone.

Gravity: The Invisible Dance That Shapes Our Universe

Imagine the universe as a grand ballroom, filled with celestial bodies swirling and twirling in an intricate dance. But what's the mysterious force that orchestrates this cosmic ballet? It's gravity, the invisible maestro that keeps everything in motion, from the smallest atom to the largest galaxy.

Why Gravity is the Ultimate Wingman

Gravity isn't just some boring force that keeps your feet on the ground. It's the ultimate wingman, playing cupid on a cosmic scale. It's the reason why the moon hangs out with Earth, why Earth is smitten with the sun, and why stars huddle together in galaxies, like cosmic partygoers.

Mass: The Heavyweight Champion of Gravity

Think of mass as the heavyweight champion of the universe. The more massive an object, the stronger its gravitational pull. It's like a cosmic magnet, attracting everything around it. That's why the sun, the heavyweight champ of our solar system, keeps all the planets in its orbit, like loyal fans circling a rockstar.

Distance: The Long-Distance Relationship of Gravity

Distance and gravity have a complicated relationship. They're like a long-distance couple, the further they are apart, the weaker their bond becomes. It's an inverse square relationship, which means if you double the distance, the gravitational force becomes four times weaker. So, even though gravity has a long reach, it gets weaker the further it stretches.

Visualizing Gravity: The Cosmic Cobweb

Imagine gravity as a giant cosmic cobweb, with invisible threads connecting everything in the universe. The closer the threads, the stronger the gravitational pull. It's like a cosmic game of tug-of-war, with massive objects pulling on each other with invisible strings.

Gravity in Action: A Cosmic Playbook

Gravity isn't just some theoretical concept; it's a force that shapes our everyday lives. It's the reason why the tides ebb and flow, why apples fall from trees, and why we don't float off into space.

Planetary Orbits: The Cosmic Waltz

The planets in our solar system are like graceful dancers, waltzing around the sun in a perfectly choreographed routine. And who's the choreographer of this cosmic dance? Gravity, of course, keeping the planets in their orbits with its invisible hand.

Tides: The Moon's Love Affair with the Oceans

The moon has a special connection with Earth's oceans, causing them to rise and fall in a rhythmic dance of tides. It's like a cosmic love affair, with the moon's gravity pulling on the oceans, creating bulges of water that we experience as tides.

Galaxy Formation: The Cosmic Huddle

Galaxies are like cosmic huddles, with stars and gas clouds clumping together under the influence of gravity. It's like a cosmic party, with gravity playing the role of the social butterfly, bringing everyone together.

Advanced Concepts: Peering Deeper into the Cosmic Mystery

Gravity is a complex and mysterious force, with many secrets yet to be revealed. But scientists are constantly exploring its depths, uncovering new and fascinating insights into its nature.

Gravitational Potential Energy: The Energy of Position

Imagine a ball perched on a hilltop, ready to roll down. It possesses gravitational potential energy, the energy of position. The higher the ball, the more energy it has, like a cosmic savings account.

Gravitational Lensing: The Cosmic Magnifying Glass

Massive objects, like galaxies, can act as cosmic magnifying glasses, bending the path of light and allowing us to see distant objects that would otherwise be invisible. It's like a cosmic zoom lens, revealing the hidden wonders of the universe.

Gravitational Waves: Ripples in the Cosmic Fabric

Imagine the universe as a giant trampoline, with massive objects creating ripples in its fabric as they move. These ripples are gravitational waves, carrying information about the most violent events in the universe, like cosmic whispers from the depths of space.

Exploring Further: The Never-Ending Quest

The study of gravity is a never-ending quest, with scientists constantly pushing the boundaries of knowledge. It's a journey of discovery, revealing the secrets of the universe and our place within it.

Conclusion: The Cosmic Dance Goes On

Gravity is the invisible force that shapes our universe, orchestrating the cosmic dance of celestial bodies. It's a force that affects us all, from the smallest atom to the largest galaxy. As we continue to explore its mysteries, we'll gain a deeper understanding of our place in the grand cosmic ballet.

The Sun's Gravitational Embrace: A Cosmic Dance of Attraction

Imagine the Sun as a cosmic dancer, a celestial maestro leading a grand ballet of planets, asteroids, and comets. Its graceful movements are dictated by an invisible force, a powerful attraction that binds the entire solar system in a harmonious embrace – gravity.

This isn't just any ordinary force; it's a love story written in the language of physics. The Sun, with its massive heart of incandescent gas, holds 99.86% of the solar system's mass. It's a heavyweight contender in the universe, and its gravitational pull is the ultimate expression of its strength.

Think of it like this: if you were to stand on the Sun's surface (ignoring the fact that it's a scorching inferno), you'd feel 28 times heavier than you do on Earth. It's like the Sun is giving you a giant cosmic bear hug!

But the Sun's love isn't limited to just the planets. It extends its warm embrace to every celestial body within its reach, from the smallest asteroid to the farthest comet. Even the Oort cloud, a distant realm of icy wanderers, feels the Sun's gentle tug.

This gravitational dance is a delicate balance of attraction and motion. The planets, like eager partners, are constantly drawn towards the Sun, but their own momentum keeps them from falling in. It's a cosmic tango, a never-ending waltz around the solar system's central star.

And just like any good dancer, the Sun knows how to use its partner's momentum to its advantage. Spacecraft, like nimble dancers themselves, use the Sun's gravity to slingshot themselves across the solar system, saving time and fuel on their journeys to distant worlds.

But even the most passionate dances must eventually come to an end. As the Sun ages, it will slowly lose some of its mass, like a dancer shedding a costume. This will weaken its gravitational embrace, causing the planets to drift slightly further away. But fear not, this cosmic ballet is far from over. The Sun's gravitational pull will continue to shape the solar system for billions of years to come.

So, the next time you gaze up at the Sun, remember that it's more than just a bright light in the sky. It's a cosmic dancer, a gravitational maestro, and the heart of our solar system, holding everything together in a beautiful, intricate dance of attraction.

The Sun's Enchanting Waltz: A Cosmic Dance of Gravity

Imagine the solar system as a grand ballroom, with the Sun as the radiant, magnetic maestro at its heart. Each planet, a celestial dancer, twirls and swirls in a captivating ballet orchestrated by the invisible force of gravity. This fundamental force, like an invisible thread, binds the Sun and its planets in an intricate cosmic dance.

Newton's Law: The Music of the Spheres

Sir Isaac Newton, a visionary physicist, composed the musical score for this cosmic ballet. His Law of Universal Gravitation reveals the harmonious relationship between mass and distance. It's a law of attraction, stating that every object in the universe, from the mighty Sun to the smallest asteroid, yearns to embrace every other object. The more massive the objects, the stronger their yearning, and the closer they are, the more intense their embrace.

Orbital Mechanics: The Choreography of the Cosmos

Picture a child whirling a ball on a string. The ball, propelled by its initial momentum, wants to fly off in a straight line, but the string, like the Sun's gravity, gently tugs it inward. This delicate balance between the ball's desire for freedom and the string's restraining embrace results in a graceful, circular motion.

Similarly, the planets, propelled by their initial formation, yearn to travel in a straight line. But the Sun's gravitational pull, like an invisible string, gently curves their paths, creating the elliptical orbits we observe. It's a cosmic balancing act, a continuous dance between inertia and gravity.

Types of Orbits: The Dance Variations

While the waltz is the most common dance in this celestial ballroom, there are variations in the choreography. Some planets, like Earth, perform a graceful waltz in a near-circular orbit. Others, like comets, opt for a more eccentric dance, swooping in close to the Sun before pirouetting away into the depths of the solar system.

Factors Affecting Orbits: The Tempo and Rhythm

The tempo and rhythm of this cosmic dance are influenced by several factors. The Sun's mass, the primary conductor of this celestial orchestra, dictates the strength of its gravitational pull. The mass of the orbiting object, the dancer's weight, also plays a role in the choreography. And the distance between the Sun and the orbiting object determines the tempo of their dance - the closer they are, the faster they twirl.

Examples and Case Studies: The Star Performers

Earth, our home planet, performs a slightly elliptical waltz, its distance from the Sun varying throughout the year. Halley's Comet, a flamboyant performer, graces the solar system with its presence every 76 years, its highly elliptical orbit taking it on a grand tour of the solar system. And artificial satellites, like the International Space Station, are human-made dancers, carefully choreographed to maintain their orbits.

Beyond the Solar System: The Grand Cosmic Opera

This cosmic ballet extends beyond our solar system, encompassing exoplanets orbiting distant stars, binary star systems performing a captivating pas de deux, and even galaxies, vast collections of stars, swirling in a grand cosmic opera.

Advanced Concepts: The Deeper Melodies

Einstein's theory of General Relativity adds a deeper layer to this cosmic music, revealing that gravity is not just a force but a curvature of spacetime caused by mass and energy. It's a melody that resonates throughout the universe, from the smallest particles to the largest structures.

Conclusion: The Eternal Dance

The Sun's gravitational pull is the invisible choreographer of this mesmerizing cosmic ballet. It's a dance that has been ongoing for billions of years and will continue for billions more, a testament to the enduring power of gravity and the beauty of the universe.

Imagine the Sun as a cosmic dancer, its graceful movements dictating the waltz of the planets. Its gravitational pull is the invisible hand that guides their every step, but this grip weakens as the dancers move further apart.

Think of Mercury, the closest planet, as a nimble ballerina, whirling around the Sun with incredible speed. Pluto, on the other hand, is a stately grand duke, gliding along at a more leisurely pace in the far corners of the ballroom. This difference in tempo is all thanks to the Sun's waning influence.

It's like a cosmic game of tug-of-war. The Sun pulls the planets in, while their own momentum tries to fling them outwards. Closer planets, feeling the Sun's powerful grip, need to move faster to avoid being pulled in. Those further away can afford to take it slow, their orbits a more languid dance.

Even comets, those celestial rebels, are bound by this invisible force. They swoop in close to the Sun, their speed increasing as they feel its fiery embrace. Then, they slingshot away, slowing down as the Sun's grip loosens, only to be drawn back again for another dizzying dance.

This intricate choreography is not just limited to our solar system. Across the vast expanse of the universe, countless stars and planets engage in their own gravitational ballets. Some stars, more massive than our Sun, command their planets with an iron fist, while others, nearing the end of their lives, gently release their planetary partners into the cosmic void.

But there's more to this story than meets the eye. Einstein's theory of general relativity adds another layer of complexity, suggesting that gravity is not just a force, but a warp in the fabric of spacetime itself. Imagine the Sun as a bowling ball on a trampoline, its weight creating a dip that causes the planets to roll around it.

And the dance is constantly evolving. The Sun, like all stars, is slowly losing mass, its fiery breath carrying particles out into space. This means its gravitational grip is gradually weakening, causing the planets to slowly spiral outwards, like dancers moving apart as the music fades.

The study of exoplanets, those distant worlds orbiting other stars, has opened up a whole new realm of possibilities. We've discovered planets locked in tight orbits around their stars, whirling around at breakneck speeds. Others, like distant ice giants, take centuries to complete a single orbit.

The decreasing strength of the Sun's gravitational field is a fundamental principle that shapes the destiny of our solar system and countless others. It's a cosmic ballet, a dance of gravity and momentum, played out on a stage of unimaginable scale. And as we continue to explore the universe, we're sure to uncover even more intricate steps in this grand cosmic waltz.

Stars and the Universe
The Sun as a star

The Sun: A Love Letter to Our Star

Forget those dry textbook descriptions! The Sun isn't just a boring ball of gas – it's a raging inferno, a cosmic artist, a life-giver with a fiery temper. It's the ultimate celestial celebrity, and we're all part of its entourage, basking in its radiant glow.

Imagine this: you're diving into the heart of a star. The core is a pulsating furnace, a nuclear mosh pit where hydrogen atoms are slam-dancing to create helium and unleashing a mind-blowing amount of energy. It's so hot here, atoms lose their cool and electrons go AWOL, creating a wild plasma soup.

Next, you're surfing outwards on a wave of light, riding the radiative zone. Photons are playing a cosmic game of pinball, bouncing around for eons before finally escaping this stellar maze.

Hold on tight! You've hit the convective zone, a bubbling cauldron of plasma where hot stuff rises, cools down, and takes a plunge back down. It's like a cosmic lava lamp, and it's responsible for the Sun's groovy, textured surface.

Finally, you reach the photosphere, the Sun's "surface" – if you can call something that's constantly shifting and changing a surface. It's a shimmering canvas where sunspots, those cool, magnetic freckles, put on a mesmerizing show.

But the real spectacle is happening above. The chromosphere, a fiery halo, dances around the Sun, while the corona, a wispy crown of superheated gas, stretches out like a celestial embrace. And don't forget the solar wind, a constant stream of charged particles that blows through the solar system like a cosmic breeze.

The Sun isn't just a pretty face; it's a powerhouse of activity. Solar flares erupt like fiery explosions, sending bursts of energy and radiation across space. Coronal mass ejections, those massive plasma burps, can send shockwaves through the solar system. It's a reminder that our star is a force to be reckoned with.

But let's not forget the Sun's softer side. It bathes our planet in life-giving light, powering the plants that feed us and driving the weather that shapes our world. It paints our skies with breathtaking sunsets and inspires poets and artists with its beauty.

Scientists are obsessed with the Sun. The Parker Solar Probe is like a daredevil, diving closer to the Sun than ever before, while the Solar Orbiter is giving us a peek at the Sun's mysterious poles. With powerful telescopes and mind-boggling simulations, we're getting to know our star better than ever.

So, the next time you feel the warmth of the Sun on your skin, remember that you're experiencing the power of a celestial wonder. It's a star with a story to tell, a story of fire, fury, and life-giving energy. It's a story that's still unfolding, and we're lucky enough to be a part of it.

Stars: The Cosmic Cauldrons

Imagine the universe as a vast, dark canvas, and stars as the celestial artists that splash it with light and color. But these aren't just static points of brilliance; they're dynamic, living furnaces, powered by the mind-boggling process of nuclear fusion. It's like each star has a tiny sun god at its core, tirelessly smashing atoms together to create the light and warmth that bathe the cosmos.

Nuclear Fusion: The Heart of the Matter

Deep within the fiery heart of a star, where temperatures and pressures reach unimaginable extremes, atoms are engaged in a cosmic dance. This is nuclear fusion – the alchemical process where lighter atomic nuclei, like shy teenagers at a school dance, overcome their initial repulsion and fuse together to form heavier nuclei.

Think of it like this: Imagine two hydrogen atoms, each with a single proton, as tiny magnets with the same charge. They naturally repel each other. But in the intense heat and pressure of a star's core, these protons are forced to get up close and personal. They overcome their repulsion and, with a little help from quantum mechanics (the magic of the universe), they "tunnel" through the energy barrier and merge, forming a new nucleus – a deuterium nucleus (a type of hydrogen with a neutron). This process releases a burst of energy, like a tiny spark in the stellar furnace.

The deuterium nucleus then partners up with another proton, creating a helium-3 nucleus (two protons and one neutron) and releasing a gamma-ray photon – a packet of pure energy. Finally, two helium-3 nuclei collide in a grand finale, forming a helium-4 nucleus (two protons and two neutrons) and spitting out two protons to start the dance all over again.

The Missing Mass: Einstein's $E=mc^2$ in Action

Now, here's where things get really interesting. The mass of the final helium-4 nucleus is slightly less than the combined mass of the four original protons. Where did that "missing" mass go? It transformed into pure energy, according to Einstein's famous equation, $E=mc^2$. This energy, released in the form of light, heat, and other radiation, is what makes stars shine and fuels the universe as we know it.

Quantum Tunneling: The Secret Ingredient

But wait, there's more! Classical physics tells us that the temperatures and pressures inside a star shouldn't be enough to overcome the repulsion between protons. So how does fusion happen at all? Enter quantum tunneling, the universe's way of bending the rules. It's like the protons have a secret passageway through the energy barrier, allowing them to sneak through and fuse together. Without this quantum trickery, stars wouldn't be able to shine.

The Life and Death of Stars: A Fusion-Powered Saga

The type and rate of nuclear fusion reactions within a star depend on its mass and age. Stars like our Sun, the "main sequence" stars, spend most of their lives fusing hydrogen into helium, like a steady, reliable engine. But as they age, they start to run out of hydrogen fuel and begin fusing helium into heavier elements like carbon and oxygen. This causes them to swell up into red giants, like cosmic balloons.

Massive stars, the rock stars of the universe, can fuse even heavier elements, creating a whole periodic table of elements within their cores. But even these giants eventually reach their limits, culminating in a spectacular supernova explosion that scatters their enriched material across the cosmos.

Stellar Nucleosynthesis: The Cosmic Cookbook

Nuclear fusion isn't just about powering stars; it's also the cosmic kitchen where all the elements heavier than hydrogen and helium are cooked up. This process, called stellar nucleosynthesis, is responsible for creating the building blocks of planets, life, and everything we see around us. Without it, the universe would be a pretty boring place, filled with just hydrogen and helium.

Examples: Stars in the Spotlight

Our Sun: Our very own Sun is a prime example of a main sequence star, steadily fusing hydrogen into helium for billions of years. It's the ultimate source of energy for our planet, driving our weather, climate, and the very existence of life.
Betelgeuse: This red supergiant, a dying star in the constellation Orion, has exhausted its hydrogen fuel and is now fusing helium into heavier elements. It's a ticking time bomb, destined to explode as a supernova in the (astronomically) near future.
Eta Carinae: This massive, unstable star system is one of the most luminous in our galaxy. It's a cosmic drama queen, prone to outbursts and tantrums, showcasing the raw power of nuclear fusion in massive stars.

Conclusion: The Cosmic Dance of Creation

Nuclear fusion is the driving force behind the stars, the celestial engines that light up the universe and create the elements that make up everything we know. It's a story of cosmic alchemy, where atoms are smashed together in a dance of creation and destruction, powered by the fundamental forces of nature. As we continue to explore the cosmos, we uncover more secrets of these stellar furnaces, gaining a deeper appreciation for the intricate processes that shape our universe.

Stars

Galaxies: Cosmic Swirls in the Starry Sea

Imagine the universe as a vast, dark ocean, sprinkled with countless islands of light. These are galaxies, each a swirling masterpiece of stars, gas, dust, and the enigmatic dark matter. They're not just pretty to look at; they're the cosmic cradles where stars are born, live, and eventually die, shaping the universe as we know it.

What's Inside These Starry Islands?

Think of a galaxy as a cosmic cocktail, with a few key ingredients:

Stars: The showstoppers, these fiery balls of gas come in all sorts of varieties, from tiny, cool red dwarfs to massive, scorching blue giants. Our own Milky Way galaxy boasts hundreds of billions of these stellar wonders.
Gas: Mostly hydrogen and helium, this is the raw material for those stellar newborns. It hangs out in clouds, some wispy and thin, others dense and ready to birth new stars.
Dust: Like cosmic glitter, these tiny particles are made of heavier elements like carbon and silicon. They may seem insignificant, but they're vital for forming planets and obscuring starlight, creating those breathtaking dark patches in space photos.
Dark Matter: The mysterious stranger at the galactic party. We can't see it, but we know it's there because of its gravitational pull on everything else. It's like the invisible glue holding the galaxy together.

A Galaxy of Shapes and Sizes

Galaxies come in a variety of styles, much like cosmic fashion:

Spiral Galaxies: Picture a pinwheel, with graceful arms spiraling out from a central bulge. These arms are bustling with gas and dust,

making them hotbeds for new star formation. Our Milky Way rocks this look.
Barred Spiral Galaxies: A bit more edgy, these have a bar-shaped structure in the center, like a cosmic bridge connecting the spiral arms.
Elliptical Galaxies: These are the elegant, older galaxies, with a smooth, oval shape. They've mostly calmed down from their star-forming frenzy and contain mostly older stars.
Irregular Galaxies: The rebels of the galactic bunch, these have no defined shape. They're often the result of galactic collisions or close encounters, leaving them a bit disheveled but full of character.

The Life and Times of a Galaxy

Galaxies have a dramatic life story, full of creation, change, and even a bit of destruction:

Birth: In the early universe, tiny clumps of matter started to come together, drawn by gravity. These clumps grew denser and hotter, eventually igniting into the first stars and galaxies.
Growth and Evolution: Galaxies are constantly changing. They make new stars, sometimes collide and merge with others, and even pull in gas from their surroundings. It's a cosmic dance of growth and transformation.
Active Galactic Nuclei (AGN): Some galaxies have a supermassive black hole at their center, like a hungry monster devouring anything that gets too close. This creates a dazzling display of energy called an active galactic nucleus (AGN), making these galaxies some of the brightest objects in the universe.

A Few Galactic Celebrities

The Milky Way: Our home sweet home, a barred spiral galaxy with a history spanning billions of years. It's part of a small group of galaxies called the Local Group, where we hang out with our galactic neighbors.
Andromeda Galaxy: Our closest major galactic neighbor, also a spiral. In about 4.5 billion years, we're scheduled for a cosmic collision and merger with Andromeda, creating a whole new galaxy. Mark your calendars!

Sombrero Galaxy: This spiral galaxy is known for its wide-brimmed "hat" – a prominent dust lane – and a bright central bulge. It's a real showstopper in the cosmic fashion scene.
Centaurus A: An elliptical galaxy with a powerful radio source at its heart. It's like a cosmic radio station, blasting out jets of particles from its central black hole.

The Future of Galaxies: A Cosmic Mystery

As the universe keeps expanding, galaxies will become more and more isolated. Star formation will wind down, and many galaxies will fade into quiet retirement. But new galaxies might still form through mergers or by pulling in leftover gas. The ultimate fate of galaxies is still a puzzle that astronomers are trying to solve.

In Conclusion

Galaxies are more than just collections of stars; they're dynamic, evolving worlds, each with its own unique story to tell. They're the building blocks of the universe, and their study helps us understand our place in the grand cosmic scheme of things. So next time you look up at the night sky, remember those swirling islands of light, and let your imagination take you on a journey through the vast cosmic sea.

Our Sun: A Tiny Spark in the Milky Way's Grand Cosmic Dance

Imagine our Sun as a dancer, twirling gracefully in an enormous ballroom filled with countless other stars. This ballroom is the Milky Way galaxy, a vast, swirling spiral of light and wonder that stretches across 100,000 light-years. Our Sun, seemingly so dominant in our solar system, is but one tiny spark in this grand cosmic dance.

A Stellar Waltz Around the Galactic Center

Located about 26,000 light-years from the Milky Way's bustling center, our Sun resides in a quieter corner known as the Orion Arm. It's like a small neighborhood nestled between the more prominent Perseus and Sagittarius arms. Here, our Sun and its planetary entourage waltz around the galactic center at a breathtaking speed of

514,000 miles per hour. This celestial waltz takes approximately 230 million years to complete – a single galactic year.

The Milky Way: A Galaxy of Surprises

Our galaxy is a barred spiral, with a central bar-shaped structure of stars, like a dazzling centerpiece in the cosmic ballroom. Spiral arms, filled with stars, gas, and dust, extend from this central bar, forming a breathtaking pinwheel shape. These arms are also home to molecular clouds, the stellar nurseries where new stars are born.

Enveloping the galactic disk is a halo, a vast, spherical region containing globular clusters – ancient families of stars that have been dancing together for billions of years. The halo also harbors a mysterious secret: dark matter, an invisible substance that makes up most of the galaxy's mass.

And at the very heart of the Milky Way lies a supermassive black hole, Sagittarius A*, a gravitational giant millions of times heavier than our Sun. This black hole's powerful pull influences the movements of stars and gas in the galactic center, like a conductor guiding the cosmic orchestra.

The Sun's Embrace: Life and Energy for Earth

Our Sun's radiant energy, in the form of sunlight, is the lifeblood of our planet. It nourishes plants through photosynthesis, providing the foundation for Earth's food chain. Sunlight also warms our world, shapes our weather patterns, and drives the water cycle.

The Sun's magnetic field, like a protective bubble, shields us from harmful cosmic radiation. However, our star can also unleash bursts of energy and particles in the form of solar flares and coronal mass ejections, causing occasional disruptions to our technology.

Exploring the Cosmic Dance: Unveiling the Mysteries

Astronomers, like curious explorers, use powerful telescopes to study our Sun and the Milky Way. Ground-based telescopes provide detailed views of the Sun's surface and atmosphere, while space-based telescopes like Hubble and James Webb offer a clearer glimpse into the Milky Way's intricate structure.

By studying our Sun and its place in the Milky Way, we gain a deeper understanding of how stars and galaxies form and evolve, and how life can arise in the vastness of the cosmos.

Examples and Case Studies

Gaia Mission: The European Space Agency's Gaia mission is like a cosmic census, mapping the positions and movements of billions of stars in the Milky Way. This data helps us understand our Sun's orbit and the distribution of dark matter.
Solar Orbiter: This mission provides close-up views of the Sun's atmosphere and magnetic field, helping us understand the origins of solar wind and space weather.
Event Horizon Telescope: This project captured the first-ever image of Sagittarius A*, confirming the existence of supermassive black holes and providing valuable insights into their properties.

Conclusion

Our Sun, the source of life and energy for Earth, is an integral part of the Milky Way's grand cosmic dance. By studying our Sun and its galactic environment, we continue to unravel the mysteries of the universe and our place within it. We are, after all, children of the stars, connected to the cosmos in ways we are only beginning to understand.

Imagine the universe as an endless ocean, and our solar system is just a tiny island in this vast expanse. Even the distances within our little island are huge compared to our everyday experiences. Now, picture venturing beyond our island, into the deep, uncharted waters of the universe. The distances become so immense that they boggle the mind. To navigate this cosmic ocean, astronomers have created special "maps" and "measuring sticks," with the light-year being one of the most important tools.

Think of light as a cosmic messenger, carrying stories from distant stars and galaxies. When we look at a star that is 100 light-years away, we are not just seeing its light; we are receiving a message that has traveled for 100 years to reach us. It's like receiving a postcard from the past, telling us what the universe was like 100 years ago.

Imagine yourself as a time traveler, gazing through a telescope at a distant galaxy billions of light-years away. You are not just seeing a galaxy; you are witnessing a snapshot of the early universe, billions of years in the past. It's like having a time machine that allows you to observe the universe's childhood.

Describe the life cycle of a star

Imagine a canvas of cosmic dust, a swirling ballet of particles dancing in the abyss. This is where stars are born, in the heart of these interstellar clouds, where the universe whispers secrets of creation.

These clouds, vast and ethereal, are like cosmic gardens, nurturing the seeds of stars. They are not simply empty spaces, but bustling metropolises of hydrogen and helium, seasoned with a pinch of heavier elements – the stardust from long-dead stars.

Picture a cloud so dense, it becomes a silhouette against the backdrop of the cosmos, a dark nebula where gravity weaves its magic. Inside, pockets of gas and dust collapse like falling dominoes, each fragment destined to become a star. This collapse is a cosmic drama, fueled by shock waves from exploding stars, collisions between clouds, and the gravitational pull of spiral galaxies.

As the cloud shrinks, it spins faster, like a pirouette on a cosmic stage. A protostar emerges, a luminous beacon in the swirling darkness. It's a hungry newborn, greedily devouring surrounding matter, growing hotter and brighter.

Imagine this protostar as a cosmic chrysalis, its outer layers shedding, revealing the radiant star within. Jets of gas erupt from its poles, like celestial fireworks announcing the birth of a new star. And around it, a swirling disk of dust and gas takes shape, a planetary nursery where worlds will eventually form.

To witness this stellar genesis, we turn our gaze to the Orion Nebula, a dazzling tapestry of gas and dust where young stars ignite. Or to the Eagle Nebula, where the iconic "Pillars of Creation" stand as testaments to the ongoing cycle of stellar birth and death.

The James Webb Space Telescope, our celestial time machine, allows us to peer into these stellar nurseries with unprecedented clarity. We unravel the mysteries of magnetic fields, the birth of massive stars, and the intricate dance between star and planet formation.

Each new discovery is a brushstroke on the canvas of our understanding, revealing the artistry of the universe and our place within its grand design. For in the birth of stars, we witness not just the creation of distant suns, but the echoes of our own origins, a reminder that we are all, in essence, stardust.

Imagine a cosmic dance floor, shimmering with stardust and echoing with the whispers of creation. In this vast ballroom, gravity takes the lead, a relentless choreographer pulling gas and dust together in a swirling waltz. As the tempo increases, these particles, once distant strangers, become intimately entwined, heating up with the passion of their embrace. This fiery courtship gives birth to a protostar, a stellar toddler still stumbling towards its destiny.

But becoming a star isn't as simple as showing up. It's a delicate balancing act, a cosmic tango between gravity's insistent pull and the outward push of radiation pressure. Think of it as a cosmic tug-of-war. On one side, gravity, the ultimate strongman, tries to crush the protostar's core. On the other side, nuclear fusion, the engine of the stars, ignites, releasing a torrent of light and heat that pushes back against gravity's relentless squeeze.

Picture a rebellious teenager, resisting the constraints of their parents. The protostar, fueled by the energy of youth, fights back against gravity's suffocating embrace. As the core temperature soars, the dance floor heats up, and the atomic nuclei within the protostar become a whirlwind of motion. They collide with increasing fervor, fusing together and releasing a surge of energy that pushes outward, like a defiant shout for freedom.

This battle between gravity and radiation pressure is a delicate dance, a constant negotiation between collapse and expansion. If gravity overpowers the outward push, the star collapses in on itself, like a deflated balloon. But if the outward pressure gains the upper hand, the star could explode, scattering its stardust across the cosmos.

The key to stellar success lies in finding the perfect balance, the sweet spot where gravity and radiation pressure hold each other in a delicate embrace. This is the moment the protostar transforms into a main sequence star, a cosmic adult ready to take its place in the universe's grand narrative.

Think of our Sun, a middle-aged star comfortably settled into its role as the life-giver of our solar system. It has mastered the art of balance, effortlessly fusing hydrogen into helium, radiating warmth and light like a contented campfire. But even our Sun, this beacon of stability, will eventually exhaust its fuel, facing a cosmic retirement that could be as gentle as a fading ember or as explosive as a supernova.

The journey from protostar to stable star is a testament to the universe's exquisite balance, a cosmic ballet where gravity and energy perform a breathtaking pas de deux. It's a story of struggle and triumph, of collapse and expansion, a reminder that even amidst the chaos of the cosmos, there exists a profound and beautiful harmony.

Imagine a star, not as a static point of light, but as a living being, breathing, evolving, and ultimately succumbing to the passage of cosmic time.

Born from the swirling chaos of a nebula, a star ignites, a radiant beacon fueled by the fiery heart of hydrogen fusion. It spends eons in this vibrant prime, a steady member of the cosmic choir, its light a testament to the enduring power of nuclear fire. Our own Sun, a middle-aged star, basks in this stage, its warmth nurturing life on our little blue planet.

But even stars cannot escape the relentless march of time. As the hydrogen fuel dwindles, the star's core contracts, a cosmic heartbeat quickening. The outer layers, like a billowing cosmic gown, expand and cool, transforming the star into a crimson giant, a stellar elder.

For some stars, this is a prelude to a spectacular demise, a supernova explosion that briefly outshines entire galaxies. But for others, like our Sun, a gentler fate awaits. The outer layers, like a whispered sigh, drift away, forming a breathtaking planetary nebula, a cosmic butterfly emerging from its stellar chrysalis.

At the heart of this celestial spectacle lies the star's remnant core, a white dwarf, a tiny ember of its former glory. This dense, Earth-sized relic, once a raging inferno, now cools and fades, a silent testament to the ephemeral nature of stellar existence.

Think of Betelgeuse, the ruby-red shoulder of Orion, a supergiant poised on the brink of a supernova, its impending explosion a cosmic drama waiting to unfold. Or consider the Ring Nebula, a delicate halo of glowing gas, the ethereal ghost of a Sun-like star, its light a poignant reminder of the star's past grandeur.

The life cycle of stars is a cosmic ballet, a dance of creation and destruction, of birth, transformation, and eventual demise. It's a story written in the language of light and gravity, a story that reminds us of the interconnectedness of all things in the vast universe, and our own fleeting existence within this grand cosmic narrative.

From the Ashes of Giants: A Cosmic Phoenix

Imagine a star, a colossal inferno, burning bright for eons. It's a cosmic titan, a red supergiant, its fiery heart a furnace where atoms are forged into heavier elements. But even titans fall. As the star's fuel dwindles, it begins a breathtaking swan song, a cosmic ballet culminating in a supernova, an explosion of unimaginable power that shatters the star and flings its stardust across the vast expanse of space.

This, however, is not the end. From the wreckage, a nebula emerges, a swirling cloud of cosmic debris, shimmering with the remnants of the fallen giant. This nebula, a stellar nursery, cradles the seeds of new stars, each one infused with the essence of its predecessor. It's a cycle of rebirth, a cosmic phoenix rising from the ashes, a testament to the universe's endless capacity for renewal.

The Supernova Spectacle: A Star's Grand Finale

Supernovae are the universe's most dramatic fireworks, explosions that briefly outshine entire galaxies. They're the grand finales of massive stars, the culmination of a stellar life cycle that began billions of years ago.

There are two main types of supernovae:

Type I: These occur in binary star systems, where a white dwarf star siphons matter from its companion until it reaches a critical mass, triggering a runaway nuclear explosion.
Type II: These are the death throes of massive stars, like our red supergiant. When these stars exhaust their nuclear fuel, their cores collapse, triggering a catastrophic explosion that shatters the star and sends its outer layers hurtling into space.

Stellar Nurseries: Where Stars Are Born

The material ejected from a supernova forms a nebula, a swirling cloud of gas and dust. This nebula, enriched with the heavy elements forged in the star's core, becomes a fertile ground for the formation of new stars.

Dense clumps of gas and dust within the nebula collapse under their own gravity, eventually becoming hot and dense enough to ignite nuclear fusion in their cores. These newborn stars, infused with the remnants of their predecessor, may have planetary systems of their own, potentially harboring life.

Examples: Echoes of Cosmic Cataclysms

The Crab Nebula: A celestial tapestry woven from the remnants of a supernova observed by Chinese astronomers in 1054 AD. At its heart lies a rapidly rotating neutron star, a pulsar, a cosmic lighthouse emitting beams of radiation that sweep across the sky.
Cassiopeia A: A young supernova remnant, a celestial ghost echoing a cataclysm that occurred just a few centuries ago. It's a powerful source of radio waves and X-rays, a testament to the immense energy released in a supernova explosion.

Supernova 1987A: A relatively recent supernova, observed in 1987 in the Large Magellan Cloud, a nearby galaxy. It provided astronomers with a front-row seat to a stellar demise, confirming the production of heavy elements in supernovae and the formation of dust in their aftermath.

The Cosmic Dance: Neutron Stars, Black Holes, and the Cycle of Rebirth

The core of a massive star that undergoes a supernova can collapse into either a neutron star or a black hole:

Neutron stars: Incredibly dense remnants, with a mass greater than the Sun packed into a sphere just a few miles across. Some neutron stars, known as pulsars, emit beams of radiation that sweep across the sky as they rotate.
Black holes: Regions of spacetime where gravity is so strong that nothing, not even light, can escape. They're the ultimate cosmic enigmas, the remnants of stars that have collapsed into oblivion.
While neutron stars and black holes don't directly participate in the formation of new stars, they play a role in the dynamics of supernova remnants and the distribution of matter in the interstellar medium.

The Everlasting Cycle: A Cosmic Symphony

The cycle of stellar life and death is a continuous process, a cosmic symphony that has been playing out for billions of years. Supernovae are the crescendos, the dramatic climaxes that enrich the interstellar medium with heavy elements and trigger the formation of new stars and planets. These new stars, in turn, will eventually reach the end of their lives and explode as supernovae, continuing the cycle and ensuring the ongoing evolution of the universe.

By studying supernovae and their remnants, we gain a deeper understanding of stellar evolution, the origin of heavy elements, and the formation of new stars and planets. It's a journey of discovery, a quest to unravel the mysteries of the cosmos and our place within it.

Our Galactic Home: A Love Letter to the Milky Way

Oh, Milky Way, you swirl of stars and stardust, your cosmic dance floor where celestial bodies waltz and twirl, you've captivated humanity since we first looked up and wondered. You're more than just a galaxy; you're our home, our birthplace, a vast canvas upon which the universe paints its most breathtaking masterpiece.

Imagine a cosmic pinwheel, a spiral of stars so grand it stretches 100,000 light-years across. That's our Milky Way, a majestic swirl of hundreds of billions of stars, each with its own story to tell. And nestled within one of those spiral arms, some 26,000 light-years from the bustling galactic center, is our little solar system, a tiny haven in this cosmic metropolis.

At the heart of our galaxy lies a supermassive black hole, Sagittarius A*, a gravitational giant that holds court over the entire Milky Way. It's a mysterious and awe-inspiring presence, a reminder of the immense forces at play in our universe.

But the Milky Way is more than just stars and black holes. It's a swirling symphony of gas and dust, the raw ingredients from which new stars are born. It's a place where ancient globular clusters, like cosmic time capsules, hold secrets of the universe's past. And it's a realm where dark matter, the universe's invisible hand, weaves its mysterious influence.

We humans, with our fleeting lifespans and limited senses, are just beginning to grasp the wonders of our galactic home. But with every new discovery, with every stunning image captured by our telescopes, our love for the Milky Way deepens.

So, the next time you find yourself under a dark, star-studded sky, take a moment to gaze upon the Milky Way. Let its hazy band of light transport you to a realm of cosmic wonders. And remember, you're not just looking at a galaxy; you're looking at home.

A Few More Cosmic Tidbits:

Gaia's Gaze: The European Space Agency's Gaia mission is like a cosmic census, mapping the positions and movements of over a billion stars in the Milky Way. It's giving us an unprecedented 3D map of our galaxy!
Dark Matter's Secrets: Scientists are on a quest to unravel the mysteries of dark matter, the invisible substance that makes up most of our galaxy's mass. Who knows what secrets it holds?
Galactic Archaeology: Like cosmic detectives, astronomers are piecing together the history of the Milky Way by studying the chemical compositions and movements of its stars. It's a galactic puzzle waiting to be solved!
Exoplanet Paradise: Thousands of planets have been discovered orbiting other stars in the Milky Way. Could some of them harbor life? The possibilities are endless!
The Milky Way is a constant source of wonder and inspiration. It's a reminder of our place in the vast universe, a cosmic family portrait that connects us to billions of stars and countless worlds. And as we continue to explore its depths, we're sure to uncover even more breathtaking secrets and awe-inspiring wonders.

Redshift: A Cosmic Symphony of Light and Motion

Imagine the universe as a vast, ever-expanding canvas, painted with the brilliant hues of distant stars and galaxies. Now, picture yourself as an art enthusiast, gazing at this masterpiece, trying to decipher its hidden stories. Redshift, my friend, is the key that unlocks these cosmic secrets, a celestial Rosetta Stone that allows us to understand the language of the universe.

Think of a jazz saxophone wailing its soulful tune. As the musician sways towards you, the notes seem to rise in pitch, becoming sharper and more urgent. But as they lean back, the melody softens, stretching out into a mellow, lower register. This, in essence, is the Doppler effect, a phenomenon that applies not only to sound waves but also to the very fabric of light itself.

When a celestial object, like a distant galaxy, hurtles away from us in the vast expanse of space, its light waves are stretched, much like the notes of our retreating saxophonist. This stretching shift the light towards the redder end of the spectrum, hence the term "redshift." Conversely, if an object zooms towards us, its light waves compress, creating a "blueshift."

Redshift, however, is more than just a cosmic color change. It's a cosmic speedometer, a yardstick that allows us to measure the mind-boggling distances between galaxies and unravel the mysteries of the universe's expansion.

Types of Redshifts: A Cosmic Trio

Just as a musical piece can have different movements, redshift comes in various forms, each with its own unique story to tell:

Cosmological Redshift: This is the grand symphony of the expanding universe, the cosmic equivalent of a majestic overture. As space itself stretches, the light waves traveling through it are also stretched, creating a redshift that increases with distance. It's like the universe is a giant balloon, and as it inflates, the galaxies painted on its surface move further apart.

Doppler Redshift: This is the solo performance of a single celestial object, like a star or galaxy, moving relative to us. If it's moving away, its light is redshifted; if it's approaching, it's blue shifted. It's like watching a dancer twirling across the stage, their movements creating a dynamic interplay of colors.

Gravitational Redshift: This is the deep, resonant bass line of a massive object, like a black hole, warping the fabric of spacetime. As light struggles to escape the intense gravitational pull, it loses energy, stretching its waves and creating a redshift. It's like a powerful singer hitting a low note that seems to vibrate through your very core.

Measuring Redshift: Decoding the Cosmic Score

Redshift is measured using a dimensionless quantity, 'z,' which represents the ratio of the change in wavelength to the original wavelength. It's like comparing the length of a stretched rubber band to its original size. The higher the value of 'z,' the greater the redshift, and the faster the object is receding from us.

Redshift and the Expanding Universe: The Big Bang's Echo

The discovery of redshift in the early 20th century was a watershed moment in astronomy, akin to hearing the first notes of a groundbreaking symphony. It revealed that the universe was not static, but rather a dynamic, ever-expanding entity, born from a cataclysmic event known as the Big Bang.

Redshift and Distance Measurement: A Cosmic Ruler

Redshift is not only a testament to the universe's expansion but also a powerful tool for measuring cosmic distances. Hubble's Law, a cornerstone of cosmology, links the redshift of a galaxy to its distance from us. It's like having a cosmic ruler that allows us to chart the vast expanse of the universe.

Redshift and the Early Universe: A Glimpse into the Past

Redshift also acts as a time machine, allowing us to peer back into the universe's infancy. Light from distant galaxies has taken billions of years to reach us, so when we observe them, we're essentially seeing them as they were billions of years ago. It's like looking at a faded photograph, capturing a moment frozen in time.

Case Studies: Redshift in Action

The Hubble Deep Field: Imagine staring at a seemingly empty patch of sky and suddenly discovering a hidden universe teeming with galaxies. That's what the Hubble Space Telescope did in 1995, revealing a treasure trove of distant galaxies, their redshifts providing evidence for the universe's vastness and ongoing expansion.

The Sloan Digital Sky Survey: This ambitious project has mapped millions of galaxies and quasars, creating a 3D map of the universe. Redshift measurements from this survey have helped refine our understanding of the universe's expansion rate and the mysterious dark energy that drives it.

The James Webb Space Telescope: This powerful new telescope is poised to revolutionize our understanding of the early universe, peering even further back in time than ever before. Redshift will be a crucial tool in its quest to unravel the mysteries of the first stars and galaxies.

Conclusion: Redshift, the Cosmic Maestro

Redshift is a fundamental concept in astronomy, a cosmic conductor that orchestrates our understanding of the universe's grand symphony. It's a testament to the universe's dynamism, a tool for measuring cosmic distances, and a time machine that allows us to glimpse into the past. As we continue to explore the cosmos, redshift will undoubtedly guide us towards new and exciting discoveries, revealing the hidden melodies of the universe.

Imagine the universe as a giant loaf of raisin bread rising in the oven. Each raisin represents a galaxy, and as the bread expands, the raisins move further apart. That's essentially what's happening in our universe!

This mind-blowing concept of an expanding universe is backed up by a fascinating phenomenon called redshift. Think of it like the Doppler effect for light. You know how a siren sounds higher pitched as it approaches and lower pitched as it moves away? Well, light from distant galaxies gets stretched as they move away from us, shifting its color towards the red end of the spectrum. The faster they retreat, the redder their light becomes.

Edwin Hubble, a rockstar astronomer, made a groundbreaking discovery in the 1920s. He realized that the farther a galaxy is, the faster it's zooming away from us. This cosmic rule, known as Hubble's Law, is like a speedometer for the universe. It helps us measure how fast it's expanding and even estimate its age!

But here's where things get even more intriguing. The expansion is actually speeding up! It's like someone cranked up the oven temperature for our cosmic raisin bread. This acceleration is driven by a mysterious force called dark energy, which makes up a whopping

68% of the universe's energy. It's like an invisible hand pushing everything apart, and we're still trying to figure out what it is.

The expanding universe has some pretty wild implications. It suggests that everything started from a tiny, super-hot point – the Big Bang – about 13.8 billion years ago. And if the expansion keeps accelerating, the universe will become increasingly cold and empty, with galaxies drifting farther and farther apart.

But hey, there's still so much we don't know! Scientists are grappling with puzzles like the Hubble Tension, a cosmic disagreement about the exact expansion rate. And the early universe, with its mind-boggling inflationary epoch, is still shrouded in mystery.

Despite these challenges, the expanding universe is one of the most awe-inspiring discoveries in science. It tells us that we're part of a dynamic, ever-evolving cosmos, and it fuels our curiosity to explore the vast unknown. Who knows what other mind-blowing secrets the universe holds?

Imagine the universe as a giant, cosmic balloon, and galaxies as colorful dots painted on its surface. As you blow into the balloon, it expands, and the dots move further apart. This is similar to how our universe is expanding, causing galaxies to move away from each other.

Redshift is like a cosmic speedometer, telling us how fast these galaxies are receding. Just as the sound of a siren changes pitch as it moves away from us, the light from distant galaxies stretches, shifting towards the red end of the spectrum. The faster a galaxy moves away, the greater its redshift.

This cosmic expansion points to a mind-boggling idea: the universe was once incredibly small and hot, like a primordial soup of energy and matter. This is the essence of the Big Bang theory, which suggests that everything we see today – stars, galaxies, planets, even us – originated from this initial explosion.

But redshift isn't just about the Doppler effect. It's also influenced by gravity, which can pull light downwards, stretching its wavelength. And even more fascinatingly, it's affected by the expansion of space itself, as if the fabric of the universe is being stretched like a rubber sheet.

The evidence for the Big Bang is like a cosmic detective story, with redshift as a key clue. The fact that almost all galaxies exhibit redshift tells us the universe is expanding in all directions. The faint afterglow of the Big Bang, known as the Cosmic Microwave Background radiation, also shows a redshift consistent with this expansion.

And just like a detective uncovers new leads, scientists are constantly refining our understanding of the universe. The discovery of dark energy, a mysterious force accelerating the expansion, has added a new layer of complexity to the cosmic puzzle.

The study of redshift is a journey into the depths of the universe, a quest to understand our origins and our place in the cosmos. It's a story that continues to unfold, with new discoveries and insights constantly reshaping our understanding of the universe and its evolution.

About Author

I am bestselling author. Data scientist. I have proven technical skills (MBA, ACCA (Knowledge Level), BBA, several Google certifications) to deliver insightful books with ten years of business experience. I have written and published 400 books as per Goodreads record.

ORCID: https://orcid.org/0009-0004-8629-830X

Azhar.sario@hotmail.co.uk

www.ingramcontent.com/pod-product-compliance
Lightning Source LLC
Chambersburg PA
CBHW031335230926
51848CB00002B/243
* 9 7 8 3 3 8 4 5 7 4 1 5 2 *